CROP PROTECTION CHEMICALS

ELLIS HORWOOD SERIES IN APPLIED SCIENCE AND INDUSTRIAL TECHNOLOGY

Series Editor: Dr D. H. SHARP, OBE, former General Secretary, Society of Chemical Industry; formerly General Secretary, Institution of Chemical Engineers; and former Technical Director, Confederation of British Industry.

This collection of books is designed to meet the needs of technologists already working in fields to be covered, and for those new to the industries concerned. The series comprises valuable works of reference for scientists and engineers in many fields, with special usefulness to technologists and entrepreneurs in developing countries.

Students of chemical engineering, industrial and applied chemistry, and related fields, will also find these books of great use, with their emphasis on the practical technology as well as theory. The authors are highly qualified chemical engineers and industrial chemists with extensive experience, who write with the authority gained from their years in industry.

Published and in active publication

PRACTICAL USES OF DIAMONDS
A. BAKON, Research Centre of Geological Technique, Warsaw, and A. SZYMANSKI, Institute of Electronic Materials Technology, Warsaw
NATURAL GLASSES
V. BOUSKA, *et al.*, Czechoslovak Society for Mineralogy & Geology, Czechoslovakia
POTTERY SCIENCE: Materials, Processes and Products
A. DINSDALE, lately Director of Research, British Ceramic Research Association
MATCHMAKING: Science, Technology and Manufacture
C. A. FINCH, Managing Director, Pentafin Associates, Chemical, Technical and Media Consultants, Stoke Mandeville, and S. RAMACHANDRAN, Senior Consultant, United Nations Industrial Development Organisation for the Match Industry
THE HOSPITAL LABORATORY: Strategy, Equipment, Management and Economics
T. B. HALES, Arrowe Park Hospital, Merseyside
OFFSHORE PETROLEUM TECHNOLOGY AND DRILLING EQUIPMENT
R. HOSIE, formerly of Robert Gordon's Institute of Technology, Aberdeen
MEASURING COLOUR
R. W. G. HUNT, Visiting Professor, The City University, London
MODERN APPLIED ENERGY CONSERVATION
Editor: K. JACQUES, University of Stirling, Scotland
CHARACTERIZATION OF FOSSIL FUEL LIQUIDS
D. W. JONES, University of Bristol
PAINT AND SURFACE COATINGS: Theory and Practice
Editor: R. LAMBOURNE, Technical Manager, INDCOLLAG (Industrial Colloid Advisory Group), Department of Physical Chemistry, University of Bristol
CROP PROTECTION CHEMICALS
B. G. LEVER, International Research and Development Planning Manager, ICI Agrochemicals
HANDBOOK OF MATERIALS HANDLING
Translated by R. G. T. LINDKVIST, MTG, Translation Editor: R. ROBINSON, Editor, *Materials Handling News*. Technical Editor: G. LUNDESJO, Rolatruc Limited
FERTILIZER TECHNOLOGY
G. C. LOWRISON, Consultant, Bradford
NON-WOVEN BONDED FABRICS
Editor: J. LUNENSCHLOSS, Institute of Textile Technology of the Rhenish-Westphalian Technical University, Aachen, and W. ALBRECHT, Wuppertal
PROFIT BY QUALITY: The Essentials of Industrial Survival
P. W. MOIR, Consultant, West Sussex
EFFICIENT BEYOND IMAGINING: CIM and Its Applications for Today's Industry
P. W. MOIR, Consultant, West Sussex
TRANSIENT SIMULATION METHODS FOR GAS NETWORKS
A. J. OSIADACZ, UMIST, Manchester

CROP PROTECTION CHEMICALS

BRIAN G. LEVER
International Research and Development Planning Manager
ICI Agrochemicals, Fernhurst, Surrey

ELLIS HORWOOD
NEW YORK LONDON TORONTO SYDNEY TOKYO SINGAPORE

First published in 1990 by
ELLIS HORWOOD LIMITED
Market Cross House, Cooper Street,
Chichester, West Sussex, PO19 1EB, England
A division of
Simon & Schuster International Group
A Paramount Communications Company

Typeset in Times by Ellis Horwood Limited
Printed and bound in Great Britain
by Hartnolls, Bodmin, Cornwall

British Library Cataloguing in Publication Data

Lever, B. G.
Crop protection chemicals.
1. Crops. Protection
I. Title
632
ISBN 0-13-194242-5

Library of Congress Cataloging-in-Publication Data available

Table of contents

Perspectives

Population growth and man's responsibility for the earth's resources.

> Be fruitful and multiply, and replenish the earth, and subdue it: and have dominion over the fish of the sea, and over the fowl of the air, and over every living thing that moveth upon the earth.
>
> *Genesis*, Chapter 1, verse 28

Food and agriculture.

> He who has bread may have many troubles, He who lacks it has only one.
>
> Byzantine proverb

> Farming looks mighty easy when your plow is your pencil and you're a thousand miles from the cornfield.
>
> Dwight D. Eisenhower

> And he gave it for his opinion that, whoever could make two ears of corn or two blades of grass to grow upon a spot of ground where only one grew before, would deserve better of mankind, and do more essential service to his country, than the whole race of politicians put together.
>
> Jonathan Swift, *Gulliver's Travels*, 1726

> He who makes two blades of grass grow in place of one renders a service to the State.
>
> Voltaire, Letter to M. Moreau, 1765

The pressures of pests.

> And the locusts went up over all the land of Egypt . . . and they did eat every herb of the land and all the fruit of the trees . . . and there remained not any green thing.
>
> *Exodus*, Chapter 10, verses 14–15

Risk from chemicals.

> Everything is a poison, nothing is a poison, it's the dose that makes the poison.
>
> Paracelsus, 16th century

Keeping pace with change.

> Now, here, you see, it takes all the running you can do to keep in the same place. If you want to get somewhere else, you must run at least twice as fast as that!
>
> Lewis Carroll, The Red Queen in *Through the Looking Glass*, 1872

The difficulty of planning ahead.

> What we anticipate seldom occurs; what we least expect generally happens.
>
> Benjamin Disraeli, 1837

Preface

Agrochemicals, as a topic, evoke a wide range of responses from different audiences. For farmers, agrochemicals are a necessary part of their system for profitable production of high quality food. For many environmentalists, they are hazardous substances which can damage wildlife and pollute soil and water. For others, they are tools like so many others, whose use carries risks but which give rise to rewards which are sufficiently great to more than outweigh those risks, provided they are correctly used.

There is no doubt that the increasing variety and use of crop protection chemicals has played a major role in the growth of agricultural productivity during the twentieth century. By reducing crop losses due to weed competition, pests and diseases, chemicals have increased the availability and reduced the cost of food for the world's expanding population, enabling most people to have a higher quality of life.

The discovery, development and marketing of agricultural chemicals has also generated profits to fund the growth of a complex science-based industry whose objectives are to exploit advances in chemistry and biology, both to meet perceived needs of agriculture and to maintain the profitable growth of the chemical industry itself.

As the use of agricultural chemicals has increased, however, so has the concern over possible adverse side-effects on spray operators, food consumers and the environment. The growth of the agricultural chemical industry has been paralleled by a growth in government regulations accelerated by the prominence of ecological lobbies and 'green' political parties, expansion of regulatory bodies and by increasingly complex toxicological and environmental testing.

The more questions science has enabled us to answer about environmental safety, the more questions there have been to ask. Answering these adds a major element to the cost of new product development and to costs of maintaining established products. The costs, risks and complexity of new agrochemical product invention and development are such that the process has become concentrated in a relatively small number of international companies aiming for worldwide markets.

Smaller companies have left the business or have been absorbed by bigger ones, being unable to finance the costs of R & D support and having insufficient economies of scale.

This book traces the development of the crop protection chemical industry and its interactions with the economic and scientific environment, with the agricultural industry and with a changing political framework. It also looks at the processes of new compound discovery, development and launch, and discusses the ways in which these have changed as the industry has become more advanced. It is intended to provide a breadth of understanding of crop protection chemicals not available from other, more specialist, books.

The first three chapters are devoted to a discussion of the scientific and economic environment within which the industry has grown. They give a broad historical account of the role of technical change in the development of agriculture and set agricultural chemicals in context with other production factors as a new agricultural technology. They then examine the development of the science of crop protection chemistry, considering the interaction between scientific advances, changing market needs and new chemical introductions.

Once the broad scene has been set, later chapters concentrate in much greater detail on the process of new product discovery and development. They consider developments in chemistry and molecular modelling, biological efficacy testing, application techniques, toxicological and environmental studies, large scale production, financial planning and product launch.

The final chapters discuss the historical growth and anatomy of the agricultural industry and speculate on the shape of things to come.

Throughout this book, the word billion is used for one thousand million (10^9).

It is hoped that the reader will gain a greater understanding of the workings of a vital but extremely complex industry and gain a new insight into part of the process of making two blades of grass grow where only one grew before.

ACKNOWLEDGEMENTS

This book would not have been possible without the immense amount of help and discussion from colleagues in ICI Agrochemicals. Particular thanks are due to Jeanette Ransom for unearthing data to show trends and support my assertions. Thanks to Ivor Darter for reading through and commenting on draft text, Dr Robin Birtley for work on the chapter on product safety, Dr Paul Worthington on synthesis, Dr John Hunt and Dr Ray Jones on production, Harry Swaine and Jeff Lawrence on formulation, David Harris on application technology, Derrick Smith on corporate and industry trends, and Dr W. G. Rathmell on future trends in biotechnology.

Thanks must also go to all those who have typed the text and to the ICI photographic library who provided the wealth of illustrations.

Thanks are also due to my family who put up with mounds of paper all over the study while I tried to assemble ideas and produce manuscript.

Finally, I would like to thank my editor, Dr David Sharp, for being so forbearing.

The views expresed in this book are those of the author and not necessarily those of ICI Agrochemicals.

1

The contribution of crop protection chemicals to agricultural development

HISTORICAL PERSPECTIVE

In a similar way to tractors, ploughs and fertilizers, crop protection chemicals have now become an integral part of the complex web of technical inputs required for modern agricultural production, and are accepted as a standard tool of the trade by farmers throughout the world. This is, however, a relatively recent development in the long history of the evolution of agriculture; a history which has paralleled the social and economic development of mankind from primitive hunters of meat and gatherers of berries to modern urban man.

A brief cameo of agricultural history sets the very recent development of high technology agriculture in context. Agricultural production has expanded through the centuries in order to meet the increasing nutritional needs of a growing world population with rising standards of living. This has been accomplished by a combination of increased efficiency in use of currently inhabited land and migration and expansion of population into virgin territory.

Prehistoric man was a hunter and gatherer of wild animals and plants, living at a bare subsistence level in a precarious ecological balance with his sources of food supply. Populations would have moved, thrived or declined in response to food availability. Our early ancestors then intensified land use in the transition from hunting and harvesting of wild fruits to deliberate cultivation of grains and vegetables and herding animals for meat and milk.

The development of agriculture and the domestication of animals revolutionized the balance between food supplies and population, giving man increasing control over his sources of food. Storable grains became the basis of the staple diet of early settled communities. Wheat and barley were first grown as cultivated crops in the Middle East in *c* 7000 B.C. Rice formed the staple food of the early Chinese civilization, as did maize for the inhabitants of Central America.

In *c* 4000 B.C. further intensification of land use was achieved through irrigation in the fertile flood plains of the Tigris–Euphrates and the Nile, generating food surpluses sufficient to release labour and skill for non-agricultural activities and the birth of early civilization.

Wooden ploughs were introduced in Egypt and Mesopotamia in *c* 3500 B.C., and by *c* 3000 B.C. draft animals were in use, further releasing labour which could contribute to the growth of non-agricultural employment and the establishment of urban communities.

By about 500 B.C. hoes and harrows were in use and yields began to improve when farmers started to fertilize soil with animal manure and grow legume crops which fixed nitrogen. These innovations set the broad pattern of agricultural inputs for the next 2000 years, until the agricultural revolution of the nineteenth and twentieth centuries.

Gradually agricultural activity spread to cover a progressively larger proportion of the cultivable land area of the old world. From the sixteenth century onwards the inhabitants of a crowded Europe further expanded their cropping area through migration to the New World, opening vast tracts of virgin territory for crop production or livestock rearing. As part of the new trade which developed between Old World and New, species of crop plant were taken from their indigenous hemisphere to other parts of the world where they could be grown well and where they became established parts of the diet. Potatoes from South America became a staple crop in Europe, Eastern Asia and China. Wheat was introduced from Europe to fill the prairies of North America, which would otherwise have remained under grass. Soybeans were introduced from China to the Americas and, more recently, from the Americas to Europe.

In some areas of abundant land but shortage of labour (notably North America), increased production was achieved by expanding the area cultivated by each agricultural worker using every available mechanical aid that ingenious inventors could develop, while in some areas of land shortage, for example in Britain, attempts were made to increase productivity per hectare through drainage, better rotations (e.g. the Norfolk four course rotation), better weed control (such as that achieved by horse hoeing and drilling in rows, developed in the early eighteenth century) and some attempts at crop selection and livestock breeding.

Land management and farm organization in Europe improved gradually from the mid-eighteenth century following the enclosure of fields and consolidation of farms, giving farmers much more control over their farms as businesses.

In 1840 Van Liebig, a German chemist, showed that inorganic chemicals could be used to replace nutrients removed by plants from the soil, but it was not until the twentieth century that the use of inorganic fertilizer became developed and accepted as a widespread practice.

Thus, until this century, the dominant contribution to expanding agricultural production to meet growing world food demands has been the expansion of the area under cropping. Improvements in transport enabled the increasing food needs of Western Europe to be met by the expansion of cereal and meat production in the Americas and Australasia, while much of Asian and African agriculture remained near subsistence level. The twentieth century has seen a technological revolution in agriculture which has made possible a rapid rate of growth of food production to feed a historically unprecedented growth of world population. This has enabled per capita food consumption to rise and has allowed a very rapid reduction in the proportion of the population engaged directly in food production, freeing them for new and different activities.

The twentieth century has seen a dramatic change in the rate of growth of world demand for food. For thousands of years, high mortality rates and relatively short life expectancy kept the rate of population growth below 0.15% per annum. Under these conditions it took 500 years for the population to double. The world population took 2 million years to reach 1 billion in the early nineteenth century. By 1900 it was 1.5 billion, 2.5 billion by 1950, 4 billion by 1975 and will be 6 billion by 2000 on the current median United Nations projections, and 11 billion or more by 2100 (Fig. 1).

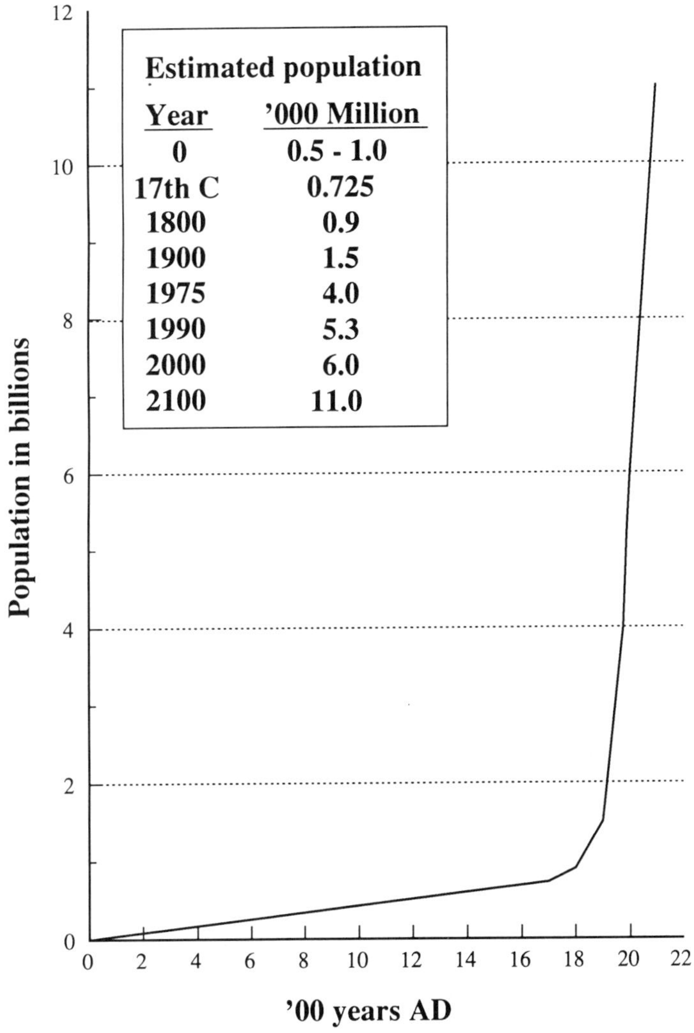

Fig. 1 — World population growth. Source: 1990 UN data and median projection.

The change in population growth rates, from the very low historical levels through a peak 1.5% in the now developed world to a current 2.4% pa in the developing world, results from a marked change in birth and mortality rates. Historically, society has existed with a balance of high birth and high mortality rates. The advent of improved nutrition and control of human diseases resulted in a decline in mortality rates while, initially, birth rates remained high, giving a rapid increase in population growth.

After a time lag, societies appear to respond to the decreased mortality rates by reducing birth rate, and a new, more stable, equilibrium is reached. Much of the developed world has passed through the demographic transition process and has a low growth rate once again. The developing world's population, however, is in a rapid expansion phase and the present high proportion of the population at pre-child-bearing age ensures that rapid expansion will continue during the remainder of this century.

To meet world demand at current living standards implies a 63% increase in cereal production, for example, between 1975 and 2000, equivalent to 2.5% per annum. The average annual grain consumption in less developed countries is *c* 180 kg per capita, which is just sufficient to meet minimum energy requirements. Average annual consumption in the USA is about 800 kg per capita (*c* 90 kg directly and *c* 700 kg via animal production as milk, meat and eggs). If global food consumption per capita increased to Western European standards, grain production would need to rise to 2.5 times the 1975 level by the year 2000.

In the developed world, per capita food consumption and nutritional standards have been rising, while the real cost of food has fallen (after adjusting for general inflation) as has the proportion of disposable income spent on food. For example, the real price of food in the USA and UK has fallen by about 50% between 1950 and 1985. The expenditure on food in the UK fell from 33% of net income in the UK in 1953 to 21% in 1984. In the USA the proportion of income spent on food fell from 70% in 1776 through 22% in 1950 to 17% in 1976. Falling costs of food have contributed to rising living standards for consumers by enabling them to spend more of their income on less essential items, for example consumer goods and leisure, and at the same time letting them enjoy higher levels of improved nutrition and more interesting and varied diets.

These consumption benefits have been made possible only by an unprecedented acceleration in the rate of technical change which has made it technically feasible and economically attractive to increase quantity and quality of food production per hectare of land. Continued growth in world food supplies must come from increased productivity.

There is estimated to be scope to expand the cropped area in the world from the current 1.5 billion ha to a maximum potential of 3.2 billion ha at the expense of grassland and forest. The growth potential lies primarily in Latin America, sub-Saharan Africa and, to some extent, the USA. Attempts to increase agricultural area by deforestation or overgrazing arid areas can, however, have dire ecological consequences. Deforestation in the Himalayan foothills has been cited as a contributory cause of increased flooding of the Indus, Ganges and Brahmaputra rivers. Cropping and overstocking on the Sahara desert fringe has resulted in a steady expansion of the desert by as much as 50 km per year in some places. Clearing of the tropical rain forests is contributing to global warming through burning of trees and destruction of a major powerhouse for conversion of carbon dioxide to oxygen and for transpiration of water. It is arguable that many areas currently under agriculture should, for ecological reasons, be returned to forestry or other conservation use, thus reducing the world area for crop production.

For many reasons, it is likely that additional land brought into cultivation would also have a lower marginal productivity than exisiting cultivated areas. The new areas will require high capital inputs and often will need technical advances to overcome many of the natural disadvantages. Expansion of cropped area is not an option open to the densely populated countries of the Old World. Continued growth in production to meet global demand projections must therefore arise from increased productivity per hectare, and some of the sources of this productivity growth are discussed later. Much of the growth in production will need to be in the developing world (because that is where the greatest population growth will occur) through adoption of both currently available and new technologies. Significant infrastructural, economic, financial and educational changes will also be needed to foster the technical change. The developing world would not be able to finance imports from the developed countries in order to satisfy its vastly increasing food supply needs, nor would it be desirable for it to have to do so.

In developed agriculture, continued improvement in productivity is still necessary to meet the changing food needs associated with rising living standards, to enable the farming community to increase its own living standards in line with those engaged in non-agricultural employment and to produce surpluses which can be exported cheaply enough to alleviate shortfalls in production in the developing world.

The rate of improvement in crop yields has, in fact, undergone a step change in the last 40 years. Fig. 2 shows the trend of wheat, barley, potato and sugar beet yields in Great Britain since about 1890.

The rate of improvement was extremely slow (less than 0.3% per annum) until the mid-1940s, when a dramatic change occurred, giving an annual growth rate in production per hectare of 2% which has been sustained for over 30 years. Similar changes in rate of yield improvement have been seen in other crops and in other countries, such as maize and cotton in the USA (Fig. 3) (each achieving a 3.3% pa yield increase since the 1930s) and rice in Japan.

Labour productivity has also risen sharply as an increasingly large proportion of the world's labour force has become employed in non-agricultural jobs (Fig. 4) and those remaining in agriculture have expected their own living standards to rise.

Labour has been replaced in many situations, and invariably in the developed world, by machinery and the use of non-renewable energy sources. This has resulted in marked increases in farm size and in the area farmed per man (illustrated in Fig. 5 for the UK and the USA).

The level of annual investment in labour-saving machinery has risen commensurately. In the USA, the annual investment in machinery and equipment rose from 300–600 million dollars (1970 constant prices) in the 1920s and 1930s to 3800 million dollars in 1978. The growth of use of farm machinery in the UK is illustrated by Fig. 6.

In the USA the growth of labour productivity has been very rapid in relation to

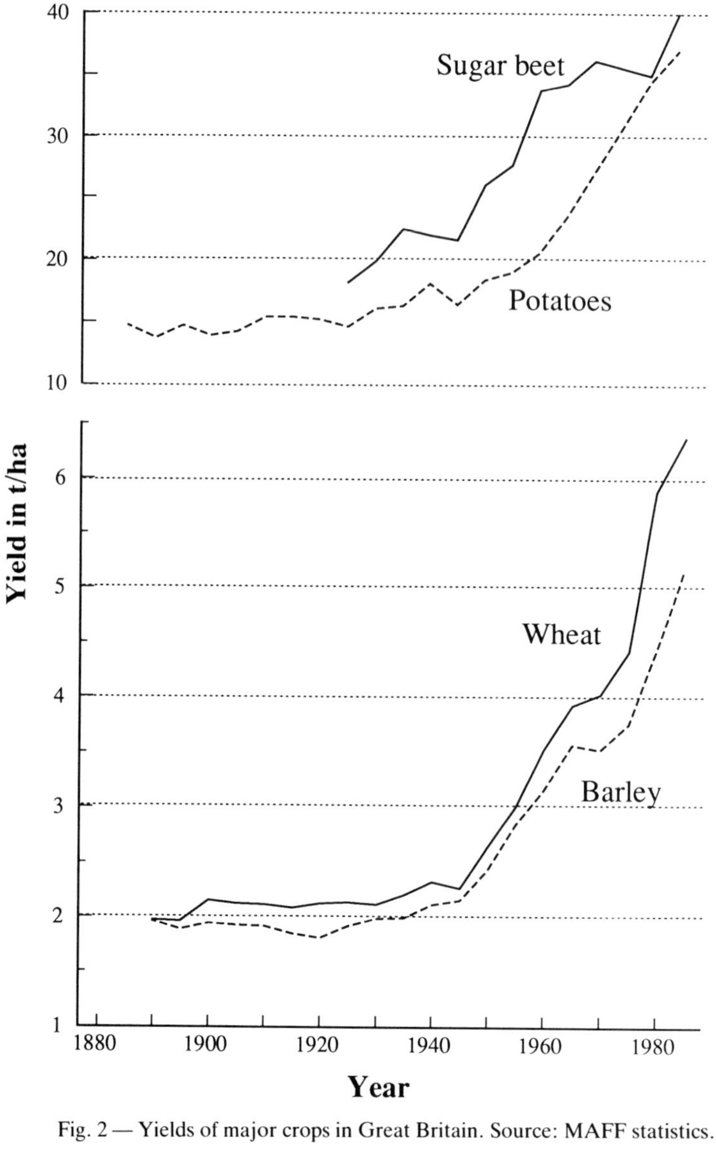

Fig. 2 — Yields of major crops in Great Britain. Source: MAFF statistics.

the growth of productivity of other inputs, for example compared to yields per hectare (Fig. 7).

SOURCES OF PRODUCTIVITY GROWTH

The sources of improvements in agricultural productivity lie in scientific and technological advances made elsewhere in the economic system and fed into

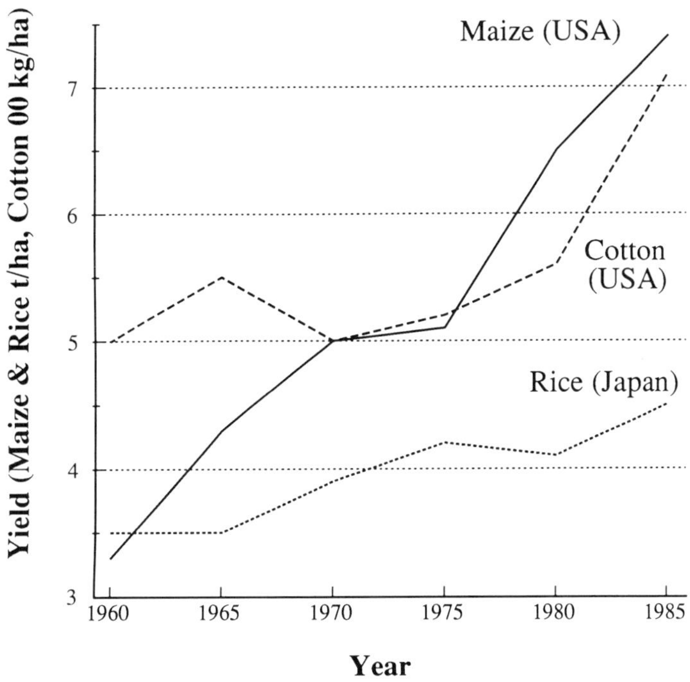

Fig. 3 — Yields of maize and cotton in USA and rice in Japan. Source: USDA and FAO statistics.

agriculture as 'inputs' which the farmer can buy: better seeds, cheaper fertilizers, new machinery, crop protection chemicals, better management techniques, and many more. A farmer buys new technology because he expects a good return on his investment. The organizations in which the new technologies are invented and developed into goods which the farmer will buy are motivated by the expectation that the sale of these 'agricultural inputs' will generate a good economic return on the resources required for invention, development, production and sale. This book is an analysis of one of the industries ancillary to agriculture whose role is the profitable generation of new technology; the agricultural chemical industry.

Before getting immersed in the detail of the agricultural chemical industry, it is, perhaps, helpful to see the role of chemical technology in the context of other contributors to agricultural productivity growth in the latter half of the twentieth century. In many ways the different fields of technology have interacted with one another, complementing each other and creating new opportunities, enhancing the economic reward.

The fundamental components of the yield and quality improvements which have been achieved are:

— plant breeding,
— plant nutrition,
— crop protection.

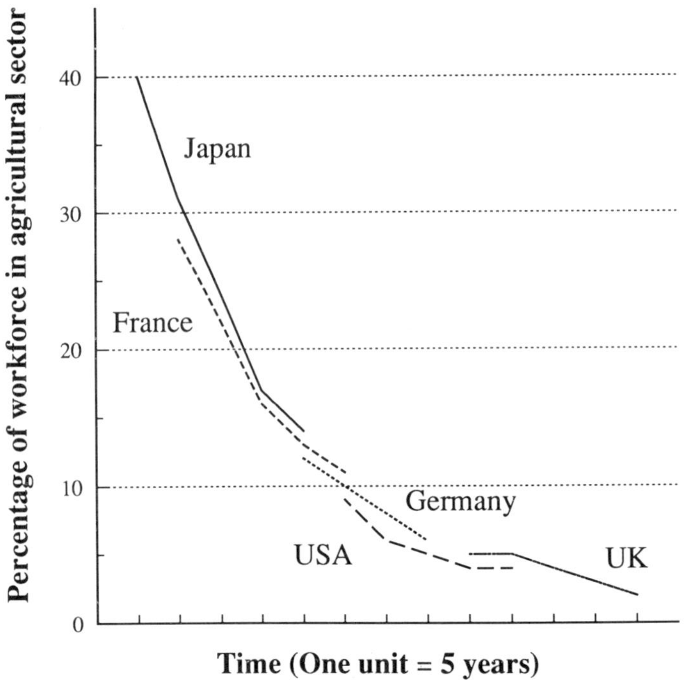

Fig. 4 — Rate of decline of agricultural workforce. Source: Presentation after Fisons in Cook (1977).

Plant breeding

The yield per hectare which a farmer is able to produce for a particular crop depends on the efficiency of the internal operation of the plant species being grown (essentially determined by its genetic make-up) and the environment in which the plant is growing (affected by natural conditions of climate and soil and also by the husbandry practices the farmer uses).

Historically, improvement in the genetic potential of crop varieties has been achieved by a slow process of selecting seed from the better crops from one season to use as the reproductive material for the next. The genetic stock has been provided by plants whose natural evolutionary trend had been towards species survival, not 'economic yield'. Economic yield is seldom the yield of the whole plant. In cereals, the economic concern is with the grain, and often with particular qualities of that grain for bread-making or malting. The leaf and stem is often a 'waste product' from farmers' points of view. In sunflower or oil seed rape, the economic yield is that of edible oils. In sugar beet, the concern is for sugar content of the root. Nature has selected for a wider range of attributes. Until the twentieth century agricultural revolution, selection of the seed from the best crops tended to imply selection of those genetic lines most suited to the natural growth conditions of low fertility, with the need for inbuilt resistance to pests and diseases, as far as this was possible and manifest in the successful crop.

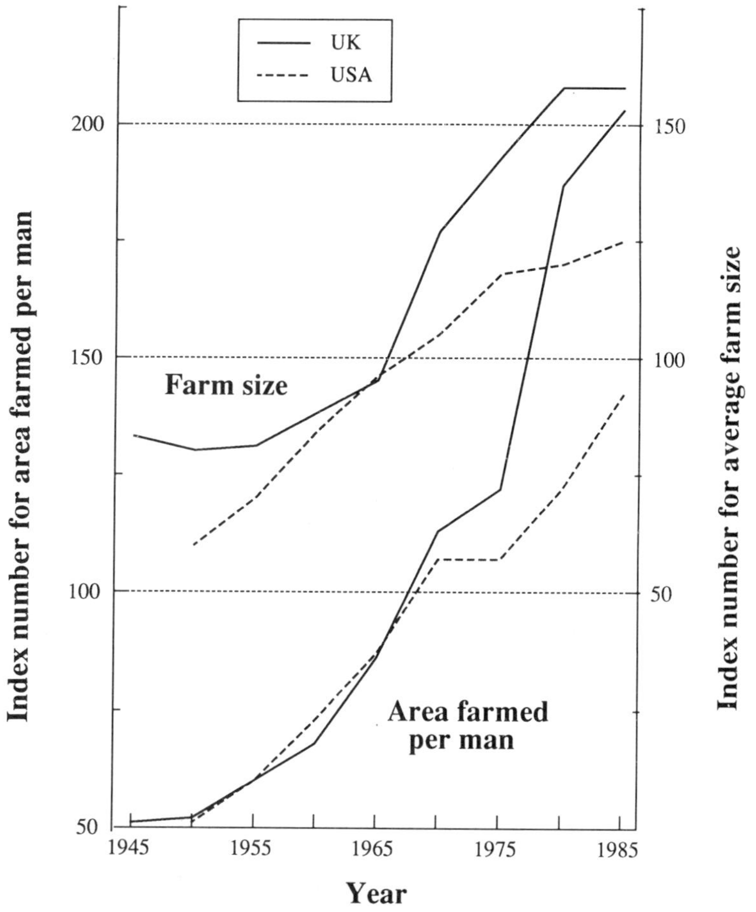

Fig. 5 — Index of growth of area farmed per man and farm size in the UK and USA. Source: USDA annual statistics and MAFF statistics.

The twentieth century has witnessed a 'yield revolution' with increased emphasis on intensive breeding programmes which have radically improved the genetic potential for yield of a wide range of crops and also developed varieties capable of giving much greater responses to other newly developed inputs such as fertilizers and agrochemicals.

The development of hybrid maize in the USA played a major role in increasing maize yields from *c* 25 bushels per acre (1.6 t/ha) in the 1930s to *c* 57 bushels per acre (3.6 t/ha) by 1960, the approximate year of full adoption by USA farmers. The Indian Council of Agricultural Research co-ordinated maize breeding scheme introduced new hybrids in the early 1960s which yielded 4–4.7 t/ha compared with 2.8 t/ha for existing local varieties.

In 1941 the Mexican Government and the Rockefeller Foundation set up a wheat

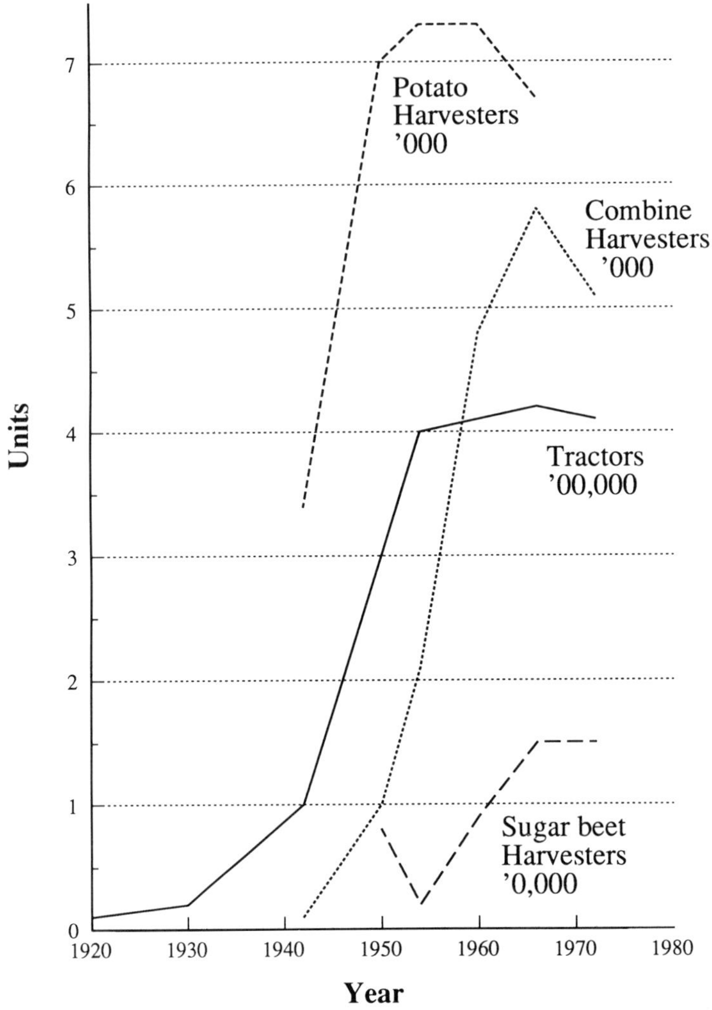

Fig. 6 — Increase in farm machinery in Great Britain. Source: MAFF (1968).

improvement programme which, following the successful crossing of Norin 10 Brevor with Mexican wheats by Dr Borlaug, was responsible for developing the dwarf and semi-dwarf and durum wheats upon which the 'Green Revolution' has been built. Mexican wheat yields rose from 0.76 tonnes per hectare in the late 1930s to 1.6 tonnes per hectare in 1958/59. Mexican stock introduced into India had given rise to wheat varieties yielding (in 1966–67 trials in Ludhiana) 4.7 t/ha compared to 2.4 t/ha average for local varieties (albeit with an extra 54 kg of nitrogen fertilizer per hectare). The Mexican wheat breeding programme merged with a comparable maize programme in 1963 to form the International Corn and Wheat Improvement Centre (CIMMYT).

The International Rice Research Institute (IRRI), established in the Philippines

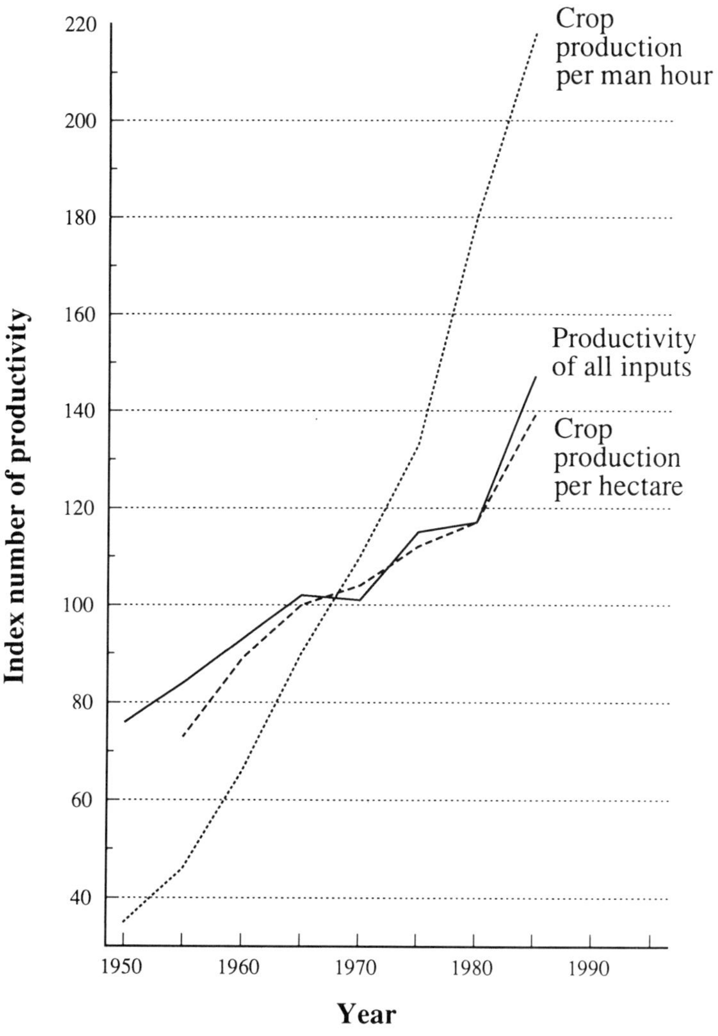

Fig. 7 — Productivity trends in USA agriculture. Source: USDA statistics.

in 1962, aimed to match the successes in maize and wheat breeding with improved rice varieties. Between 1962 and 1975 the Institute produced 11 varieties of international importance and other high yielding lines have been developed by other agencies.

Breeding programmes have not only improved yields but in many situations improved the productivity of a cropping programme by shortening maturation time and allowing multiple cropping. Many new rice varieties have 95–130-day maturation periods compared with a traditional 150 days and are not less sensitive to day-length. This allowed, in many areas, an additional crop to be grown each year. Shorter duration wheat varieties have fitted more easily into multiple cropping patterns in India than the traditional day-length sensitive, late maturing tall varieties.

Plant nutrition

Considerable improvements have also been achieved in crop husbandry — the creation of an artificial and more beneficial environment for crop growth. Early improvements came with the establishment of irrigation systems which reduced the constraint on growth which was imposed by water stress.

Another limiting factor was nutrient supply. Early solutions included the use of farmyard manure, bone meal, and beneficial rotational crops. For example, nitrogen-fixing legume crops were commonly grown as part of an arable rotation to fix nitrogen and improve the yield of succeeding cereal crops.

Although Van Liebig had demonstrated in 1840 that inorganic chemicals could be used to replace nutrients removed by plants from the soil, it was the end of the century before inorganic fertilizers came into widespread use.

During the twentieth century, fertilizer use has grown steadily worldwide (Fig. 8), reducing limitations imposed by a shortage of plant nutrients.

During the period 1950 to 1980 fertilizer use per hectare of crop increased by 10.5% per annum in Africa, 13% per annum in the Far East, 12% per annum in Latin America, 6.2% per annum in North America, 5% per annum in West Europe and 5% per annum in Japan. Rates of use have reached an average 200 kg nutrients per hectare per year in Europe and 430 kg/ha in Japan. For each crop variety, husbandry programme and location, there is, however, a diminishing response of yield to extra fertilizer applications. The response is limited by the genetic ability of the plant to use the additional nutrients in an economically useful way and by the existence of other environmental limitations. Surveys of fertilizer use in the UK have indicated that actual usage of nitrogen by the mid-1970s was close to the technical (and economic) optimum on cereals and slightly above the optimum on sugar beet. (Subsequent development of cereal fungicides and straw shorteners has since raised the economic optimum.) Economic optimum levels for fertilizer use exist for all husbandry systems, crops and locations. These are determined by the physical response of yield to nutrients, the costs of nutrients and the prices received for the crop. As will be discussed later, the response can be changed through crop breeding and by alteration of other aspects of husbandry. Fig. 8 suggests, however, a slowing of the rate of growth of fertilizer consumption per hectare in the more developed agricultures (notably Europe and North America), indicating that usage is generally approaching, or has reached, the economic optimum in those countries.

Crop protection chemicals

Plant breeding has tried to raise yield potential while the use of irrigation and fertilizers has sought to exploit that potential. Considerable advances have also been made in techniques for protecting yields from the ravages of weeds, insects and diseases. The importance of weed competition and of losses due to insects have been known since antiquity and the nature of fungal parasitology since the mid-1850s. Weeds have been controlled by hand or with farm machinery for centuries, but effective means of controlling insects and fungal pathogens are of very recent origin. The use of sulphur to control vine powdery mildew began in 1848 and the value of Bordeaux mixture for vine downy mildew control was discovered in 1882. Paris Green (a compound of copper acetate and copper arsenite) was introduced in 1869 to

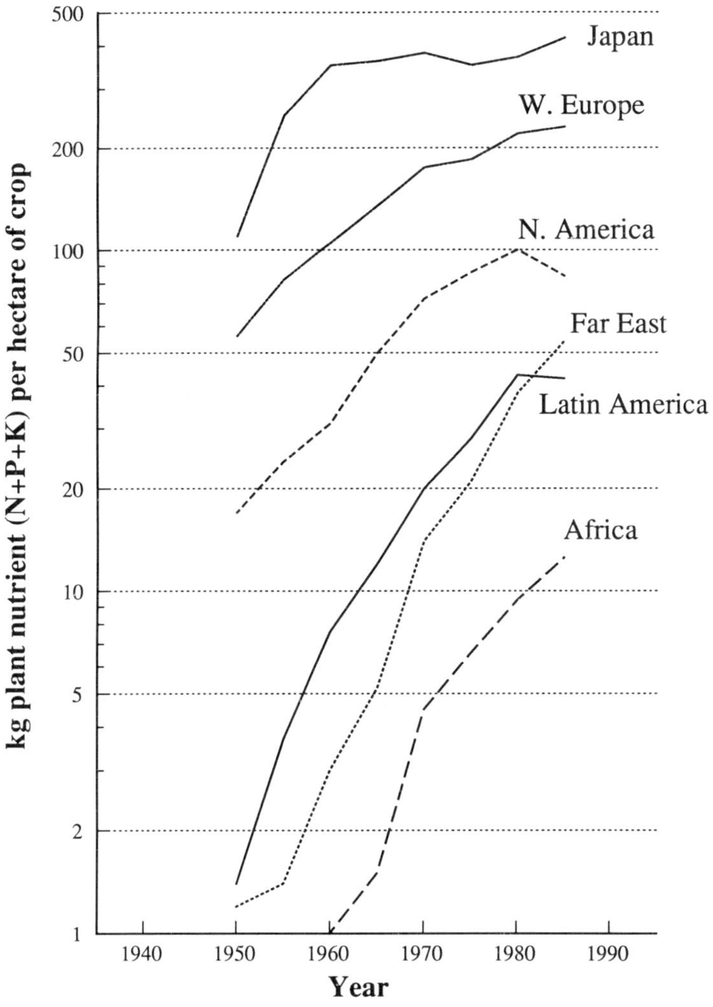

Fig. 8 — Growth of fertilizer use by world region as total kilograms of major plant nutrients (nitrogen, phosphate and potash) per crop hectare. Source: FAO.

control the spread of Colorado beetle. Lead arsenate made its debut in 1899 for the control of a gypsy moth plague in the USA.

The first chemical noted to behave as a selective herbicide was copper sulphate. Bonnet, a French vinegrower, discovered in 1897 that copper sulphate, which he had used as a fungicide, would kill charlock (*Sinapis arvensis*) without affecting grass. However, the period of rapid growth of chemicals for crop protection did not begin until the 1940s with chemical inventions of such importance as the insecticides BHC and DDT and the phenoxy-acid hormone herbicides, 2,4-D and MCPA†.

† 2,4-dichlorophenoxyacetic acid and 2-methyl-4-chlorophenoxyacetic acid.

Since then, many major chemical companies have intensified their search for new crop protection chemicals, increasing rapidly the number of patents, the number of products introduced into the market (Fig. 9) and the total value of the crop

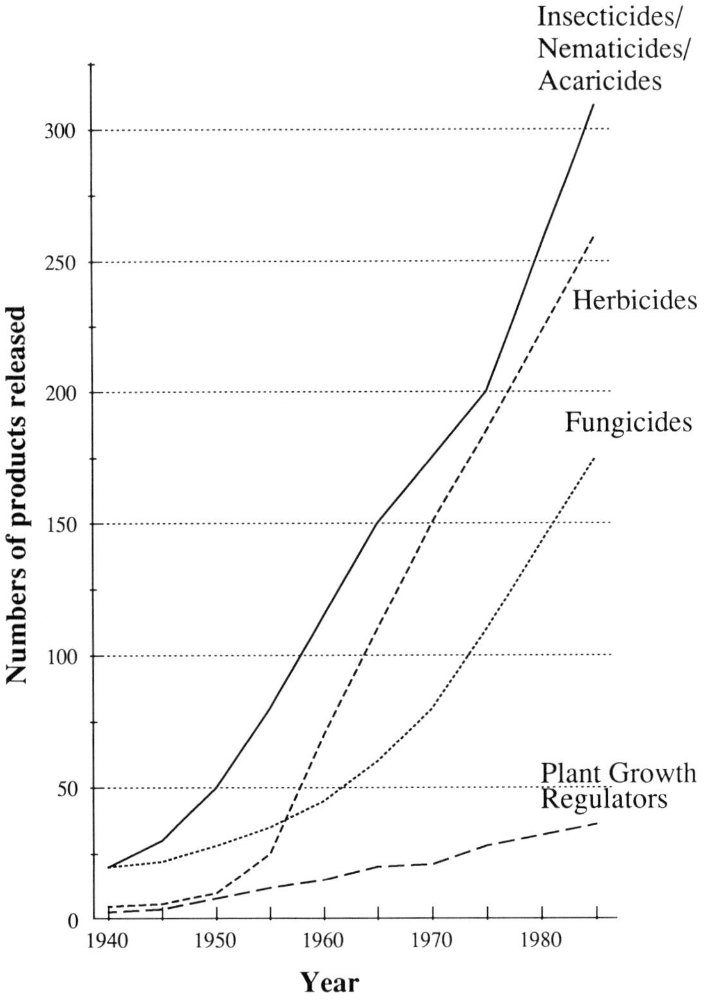

Fig. 9—Cumulative number of crop protection chemicals introduced on to the market. Source: Worthing and Walker (1987).

protection chemicals business which rose from $1 bn in 1950 to about $20 bn in 1990, at 1986 constant prices.

Products now exist to combat most of the major weeds, pests and diseases of crops, and their use has been widely adopted by farmers throughout the developed world and increasingly in the developing world. Adoption curves for herbicides in the USA and UK are illustrated by Figs 10 and 11.

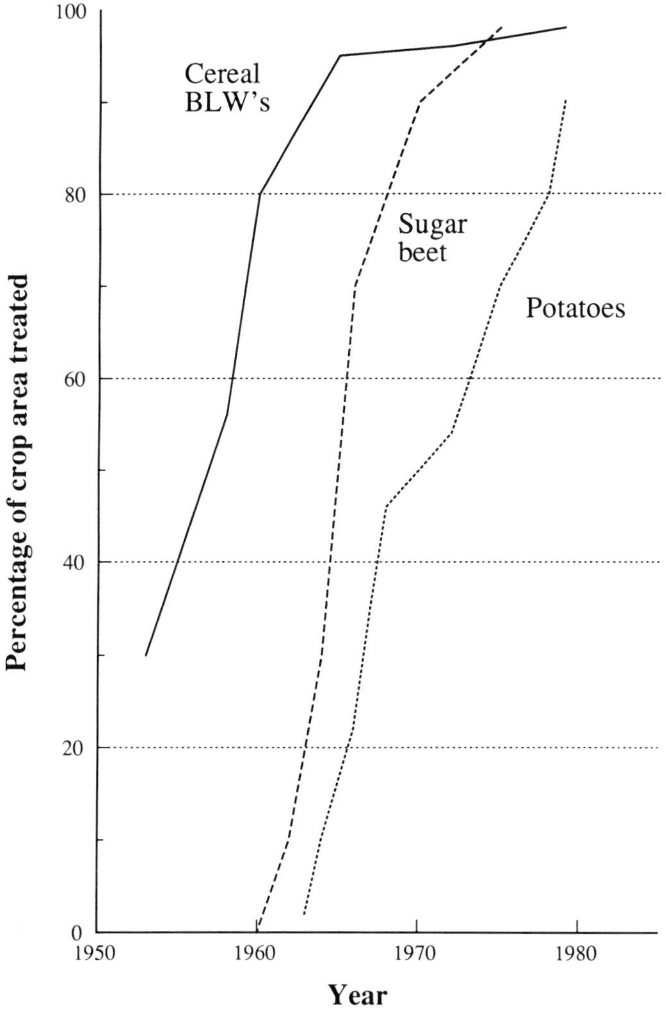

Fig. 10 — Percentage of the area of major crops treated with herbicides in the UK (BLWs=broad-leaved weeds). Source: ICI data.

The economic value of pesticide use is extremely variable from crop to crop, from pest to pest, and according to degrees of infestation. In most situations where chemicals are used, the average benefit to the user in increased yield is probably worth at least three times the cost treatment. In the extreme, the use of pesticides may be essential if the crop is to be grown at all. For example, coffee rust disease

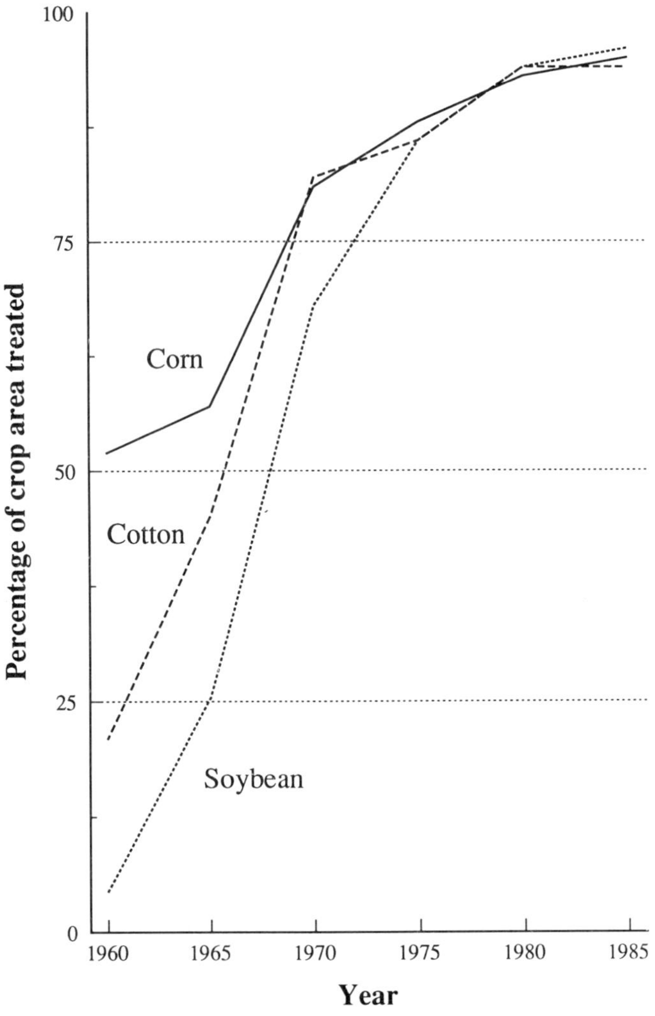

Fig. 11 — Percentage of the area of major crops treated with herbicides in the USA. Source: Doane Market Surveys, USDA.

destroyed the Sri Lankan coffee industry between 1870 and 1880 and the economic production of coffee in India and Africa is now dependent upon use of fungicides.

Many estimates have been made of the economic impact of pesticides. In a series of experiments between 1943 and 1947 Blackman and Roberts obtained cereal yield increases of over 20% as a result of the use of selective herbicides. Interestingly, the level of broad-leaved weed infestation in the typical UK cereal crop appeared to have been reduced sufficiently by the mid-1960s as a result of the repeated use of weedkillers for a survey of sprayed and unsprayed cereals by Evans in 1969 to show no significant overall benefit from broad-leaved weed control. Following selection pressure, grass weeds gradually came to be the dominant problem and yield increases of 3–148% were recorded as a result of wild oat control.

The benefits from mildew control on spring barley in the early 1970s were estimated to be worth £14 per hectare (from an average 6.5% yield increase) for a treatment cost of £4.6 per hectare — a benefit:cost ratio of 3:1 (at 1972 prices). The use of newer broad-spectrum fungicidal mixtures, including triazole fungicides, has increased the yield of wheat in France by over 15%, worth £75 per hectare for a treatment cost of £18 per hectare, giving a benefit:cost ratio of 4:1. The use of fungicides on cereals throughout Western Europe probably results in an extra 2–3 Mt cereals per year, worth £200/£300m for an expenditure on chemicals of about £100m.

The use of inorganic insecticides on cotton in the USA against boll weevil (*Anthonomus grandis*) and bollworm (*Heliothis* sp) between 1928 and 1946 gave about a 25% increase in yield over untreated crops. The newer products used between 1947 and 1958 gave about 50% increase. Data from 1976–1978 university trials for the first synthetic pyrethroid compounds indicated increases of 150%, worth $617 per hectare for an expenditure of $89 per hectare (a 7:1 benefit:cost ratio), compared with yield increases of 120% for standard methyl parathion/toxaphene/chlorphenomodone mixtures.

In Japan, comparisons between rice plots receiving standard recommended pesticide programmes and untreated plots over a ten-year study period showed an average yield of 5.8 t/ha hulled rice on treated and 2.7 t/ha on untreated plots.

In 1972, Headley estimated that on average in the USA, all pesticide usage generated four dollars of added benefit per dollar of pesticide cost.

The macro-economic benefits of pesticide use reflect the benefits to individual farmers. A USA study in 1971 by Shaw estimated that, without pesticide use, USA crop output would be reduced by 30% and prices increased by between 50% and 70%. Exact estimates are very difficult to make but food prices are very sensitive to the balance between supply and demand.

Interactions between technologies

There has been a very marked interaction between the various technologies. For any single new input alone, there is a diminishing return to its incremental use. As mentioned earlier, the response to additional fertilizer, under conditions in which other husbandry practices are constant, rapidly reaches a plateau. This response curve can be changed significantly by developing a new variety or adding further complementary inputs such as herbicides or fungicides.

In fact, the breeding programmes for new varieties have often sought deliberately to produce genotypes which are more responsive to inputs. Traditional *indica* rice varieties, for example, are adapted to the low light intensity of the tropical rainy season and to low levels of soil fertility. Application of fertilizer to these varieties results in excessive vegetative growth — excessive height, tillering and leaf area — fewer spikelets, a decrease in the proportion of grain to straw and, in some circumstances, a reduction in yield.

Historical selection has favoured these strains. Contrastingly, breeding programmes have sought to create varieties which respond to extra fertilizer by producing extra grain rather than vegetative growth, which are resistant to lodging and have a shorter growing period to allow multiple cropping. A considerable body of fertilizer response data has been generated by the IRRI for new and traditional rice varieties, and exemplary fertilizer response curves are illustrated by Fig. 12. Similar data exist for yield responses to fertilizer for most crops.

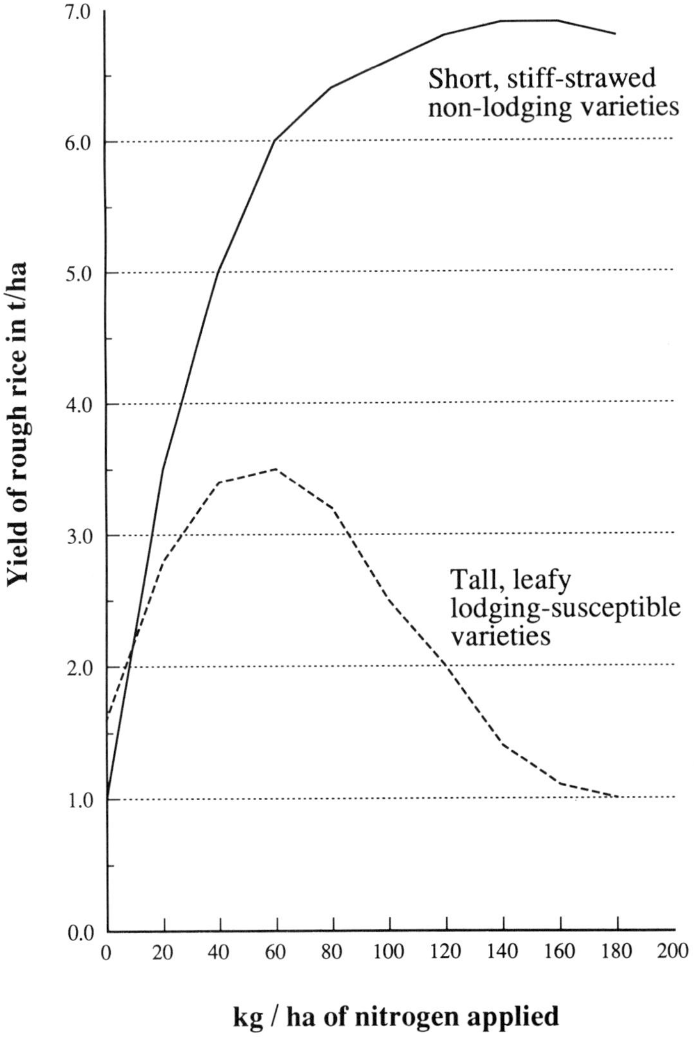

Fig. 12 — Relationship between rice variety and response to nitrogen. Source: IRRI, data in de Geus (1973).

Increased soil fertility also increases weed growth, with consequent yield-depressing effects due to competition. Fig. 13, again based on IRRI data, illustrates

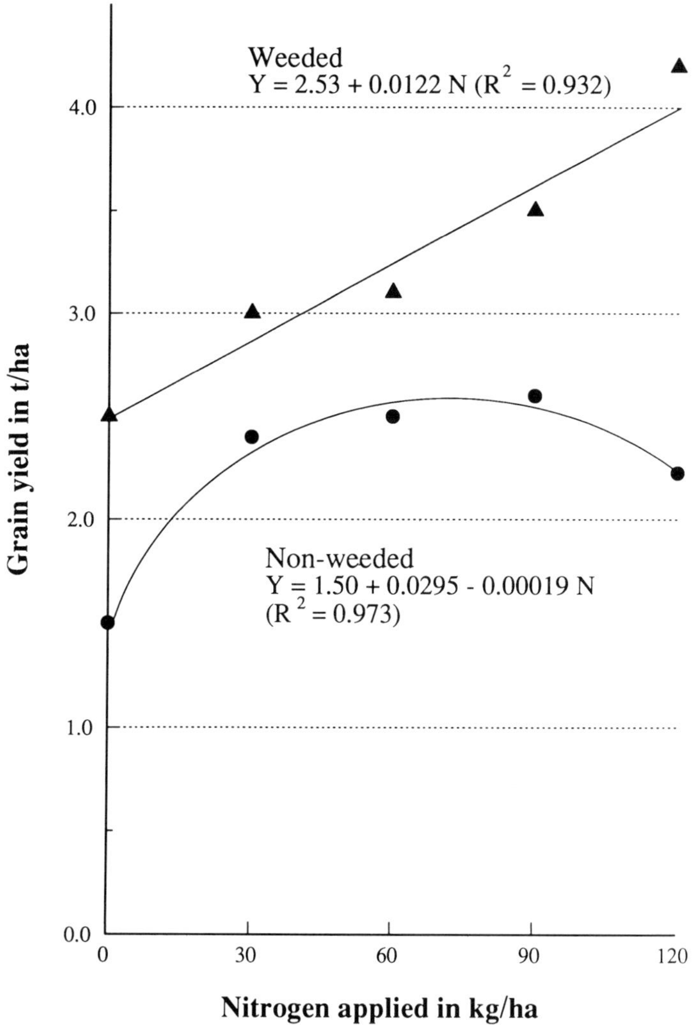

Fig. 13 — Yield response interaction between fertilizer treatment and weed control in rice at IRRI in the 1973 wet season. Source: IRRI (1974).

the interaction between weed control and fertilizer use in determining yield. Numerous other examples of the multiple interaction between new technologies are available.

Early work with insecticides in cotton in the Sudan produced data on the interaction between nitrogen fertilizer and insecticide use (Table 1).

In both years the yield benefit due to the combination of improved nitrogen availability and broad spectrum insect control (jassids and whitefly) was considerably greater than the sum of benefits resulting from either technology on its own.

In the developed world the rate of yield advance to be achieved by existing

Table 1 — Yields of seed cotton in factorial experiments in the Sudan, Gexira (Proctor, 1974)

	Yield/ha (percentage increase over control)			
	1961/1962		1962/1963	
	With N	Without N	With N	Without N
With endrin sprays	2.63 (25%)	2.22 (5%)	1.30 (59%)	0.97 (18%)
Without endrin	2.37 (12%)	2.11	0.90 (10%)	0.82
Standard error	+0.10		+0.04	

technologies is slowing down as 'best practice' is progressively more widely adopted. If further yield increases are to be sought in the future, new technologies will be needed. Possible sources for continued advance lie in chemicals to regulate plant growth or in faster breeding through advances in biotechnology.

Improvements in quality and value of production

Rising living standards have also generated a greater demand for food of higher quality and for reduced seasonality of supply, which carry their own demands for altered husbandry practices.

The demand for uniform, blemish-free apples, for example, has necessitated a comprehensive programme for insecticide and fungicide sprays to prevent maggot damage and skin blemishes caused by apple scab (*Venturia inaequalis*). The dramatic effect of a pesticide programme on marketable yield of apples is illustrated by Table 2.

Table 2 — Marketable yield of two apple varieties in West Germany

	Cox's Orange Pippin			Golden Delicious		
	Yield t/ha	Percentage marketable	Marketable yield t/ha	Yield t/ha	Percentage marketable	Marketable yield t/ha
'Normal' insecticide, acaricide and fungicide treatments	16.2	85	13.8	25.6	80	20.5
Untreated	9.5	35	3.3	16.7	25	4.2

The development of the canning and frozen food industries has resulted in the need for crops free of any pest, weed or disease. Any insects, weed seeds or blemished produce in a packet or tin of vegetables is totally unacceptable to consumers and would result in the produce being rejected.

PLANT GROWTH REGULATORS AS A NEW SOURCE OF PRODUCTIVITY GROWTH

The twentieth century has seen an unprecedented growth in agricultural productivity linked with advances in chemistry and biochemistry, but there is still enormous scope for further advance. Plant form and function have, so far, been modified by man only to the extent that available genetic material will allow. Long and complex breeding programmes have led to combinations of existing genetic material which serve man's needs better than those commonly found in nature and which have been selected by evolution to help 'survival in the wild'.

However, chemical and biochemical science are now beginning to have an impact here. As discussed earlier in this chapter, the first agricultural effects chemicals were simple inorganic compounds to supplement natural sources of nutrients, notably nitrogen (supplied as simple nitrates or ammonium salts) phosphorus (as simple phosphates) and potassium (as simple potassium salts). In their earlier forms these inorganic compounds were supplied to agriculture from mined naturally occurring sources.

The first chemicals to control pests and diseases were, again, simple inorganic substances, such as sulphur, copper sulphate, lime and lead arsenate. More complex organic molecules were then synthezised and shown to have valuable and specific biological properties. Their major contributions have, however, been in selectively killing unwanted organisms; in some ways a rather crude 'all or nothing' biological effect.

A relatively new and unexplored area of opportunity lies in the application of chemical technology to modify crop plants in an economically beneficial way. This is a more subtle requirement, and there are two alternative approaches. The furthest developed is the use of chemical treatments as plant growth regulators to modify or block biochemical processes which would otherwise occur, driven by genotype and environmental conditions. The way is also opening to the use of biochemical techniques directly to modify plant genotypes through creating new genetic material, and thereby beneficially altering plant form and function.

The idea of using chemicals to improve the growth and development of crop plants is not new. The possibility first arose in the 1930s with the discovery of natural plant hormones; the auxins. Synthetic indol-3-yl-acetic acid (IAA) first became available for horticultural use in 1936, to be followed by naphthyl acetic acid (NAA) and, in the 1940s, by the herbicidal auxins 2,4-dichlorophenoxy acetic acid (2,4-D) and 4-chloro-2-methyl phenoxy acetic acid (MCPA). The gibberellins were recognized as natural plant hormones in the early 1940s. Some of the economic benefits which can be achieved by using plant growth regulators are illustrated by the following examples.

Uses of plant growth regulators to increase value of production:

(a) Yield increase
- — Chlormequat (CCC) to shorten and stiffen wheat straw and paclobutrazol to shorten and stiffen rice straw, allowing increased use of nitrogen fertilizers, to produce higher yields and also making harvesting easier by reducing lodging (the tendency for crops to fall over).
- — Gibberellic acid (GA_3) to increase the fruit set of mandarins, clementines, tangerines and pears.
- — GA_3 to overcome losses of apple yield due to frost damage.
- — GA_3 to increase the berry size of seedless grapes, often taking the fruit into the fresh rather than the raisin market.
- — GA_3 to overcome low temperature constraints to sugar cane growth in Hawaii.
- — Ethephon to stimulate latex flow in rubber.
- — Glyphosine and glyphosate to ripen sugar cane.
- — Paclobutrazol to increase yields of tropical fruit such as mango and avocado.

(b) Quality improvement
- — GA_3 coupled with mechanical thinning to increase the berry size of seedless grapes in California to give a premium table product.
- — Ethylene to de-green citrus.
- — GA_3 to reduce the incidence of skin creasing of Valencia oranges (a physiological rind disorder which renders the fruit unacceptable for export).
- — Daminozide, chlorphonium chloride or paclobutrazol to 'dwarf' pot plants to make them more acceptable to the consumer market for house plants.
- — GA_3 and $GA_{4/7}$ to delay ripening and improve the storage life of bananas.
- — $GA_{4/7}$ mixed with benzyl adanine to improve the shape of Red Delicious apples in the USA.

(c) Value increase

- — GA_3 to advance or retard maturity of globe artichokes in order to capture higher prices outside the main production season.
- — GA_3 to enable storage of grapefruit on the tree to extend the harvesting period and capture higher priced markets.
- — Paclobutrazol to increase size and advance harvesting of peaches.

(d) Extend season and increase continuation of supply

- — Maleic hydrazide to extend the storage period for bulbs and root crops.
- — GA to force rhubarb for early production.

Cost saving uses:

(a) Direct Labour Replacement

- — Maleic hydrazide and fatty alcohol contact bud killers to control the growth of suckers in tobacco.

— Naphthyl acetic acid to thin overset fruit.
— Paclobutrazol to reduce pruning requirements of fruit trees.

(b) Facilitating mechnanization

— Fruit looseners (e.g. ethephon) to aid synchronized ripening and loosening of fruit for mechanical harvesting.
— Desiccants and defoliants to aid harvesting of cotton, legume seed crops and other oil seed crops.
— Retardants (e.g. daminozide) to reduce the length of peanut runners and facilitate harvesting.
— Retardants (e.g. mepiquat chloride) to manage the growth of cotton, aiding insecticide programmes and harvesting.

(c) Simplifying plant breeding

— A gibberellin A_3/A_4 mixture with naphthyl acetic acid is being developed to encourage precocious flowering of conifers (1 year old as opposed to 12–15 years old) to facilitate breeding programmes.

Many of the existing uses of plant growth regulators are highly profitable to growers. The use of GA_3 to increase the set of mandarin oranges, for example, can give a benefit/cost ratio of 9:1. Use of GA_3 to make seedless grapes in Chile suitable for export to the USA fresh market, rather than local raisin production, can give a 40:1 benefit/cost ratio and a more than six-fold increase in crop value per hectare.

In Hawaiian sugar cane, growth regulators have been shown to improve many aspects of crop production. Maleic hydrazide, diuron and diquat reduce or prevent flowering, an effective suppression of flower tassel formation giving a 15% increase in yield (equivalent to 3.25 tonnes of sucrose per hectare). Gibberellic acid applied during the colder winter months of the growing season has been shown to give a gain of 0.6–1.5 t/ha in sugar. Chemical ripeners, e.g. glyphosine, can add a further 10–15% to the sucrose yield.

Despite their diversity and importance within particular cultural systems, virtually all of the uses described above, with the exception of compounds such as CCC (chlormequat chloride) to reduce lodging on cereals, are specialist techniques and are of small commercial significance in terms of quantity of chemical used. The use of cereal anti-lodging compounds on cereals has now become very widespread in the intensive, high input/high yield agricultural systems of Northwest Europe. Even so, none of the growth regulators which have been developed during the last 20 years are of comparable market value to a major herbicide. The total value of the current market for growth regulatory chemicals is an order of magnitude smaller than that for the major groups of pesticides and is expected to remain so for the immediate future (Table 3).

For plant growth regulators to make the scale of impact on agricultural productivity that pesticides have achieved, they must find application on major crops and must be sufficiently cost-effective and reliable to generate benefit:cost ratios of a similar order to pesticides, in excess of 3:1. One of the early goals of agricultural chemicals research was to increase the yield of cereals by using plant hormones, and

Table 3 — Estimated value of sales for the major agrochemical effects ($ m) (source: Wood Mackenzie)

	$bn 1987
Herbicides	8.6
Insecticides	6.1
Fungicides	4.1
Desiccants	0.3
Plant growth regulators	0.4
Other	0.5
Total	20.0

it was the work on auxins with this aim in mind, during the early 1940s, which led to the unexpected discovery of the value of the phenoxy-acid compounds (notably 2,4-D and MCPA) as selective herbicides. The use of chemicals to increase yields of major arable crops has remained a target for many researchers ever since. Some near successes have been achieved, for example the use of TIBA (triiodobenzoic acid) to increase the yield of soyabeans was commercialized in the USA but it proved too unreliable under practical conditions.

Intensification of research effort into plant growth regulation may bring the type of breakthrough required, but it is far from clear where the greatest chance of technical success lies within the large range of possible avenues of work, which include, for example, stimulation of photosynthesis *per se*, reduction of photorespiration in C3 plants, better crop canopy management to use light receipts more effectively over the growing season, improved assimilate transport and sink : source balance, enhanced nitrogen fixation in legumes and many others.

Further speculation on future developments in agricultural chemistry and biochemistry for conventional areas of application for pest control, for plant growth regulators and for genetic manipulation will be discussed in Chapter 10.

The next chapters, however, concentrate on the evolution of chemical solutions to the problem of weeds, insects and fungal diseases.

2

The pest invasion

AN OPEN ECOLOGICAL NICHE

As man has attempted to tame nature by selecting and adapting plants to suit his needs, and as he has created an artificial agricultural environment in which to grow his crops, so he appears to have created great opportunities for nature to strike back.

The intensification of agricultural production and the establishment of monocultures of particular agricultural crops, often with a relatively narrow genetic base selected for yield performance, have provided an attractive environment for the development and spread of weed, insect and fungal pests for whom the lush crops are ideal hosts. As the following sections show, a number of insect and fungal pests have exploited the ecological niches created by agricultural practice with devastating consequences.

Weeds become a problem almost by definition. Once one wants to produce a pure crop stand of any particular species, any other type of plant which intrudes is, to some extent, undesirable and thereby becomes a weed. Each intruder will, to a greater or lesser extent, compete with the crop plants for light, moisture and nutrients and, in so doing, will reduce the yield of crop. Weeds can also impede harvesting and may contaminate the produce with their seeds.

Since the earliest times, weed problems have beset crop production but have been dealt with, as far as possible, by mechanical means. This may have been by hand, with hand-operated tools or by powered tools, from the horse hoes of the early eighteenth century to the tractor drawn implements of today.

THE MARCH OF THE FUNGI†

Fungal diseases have been known since antiquity, with references to 'blasting and mildew' in the books of Deuteronomy and Amos. (Deuteronomy 28, 22 'the Lord shall smite thee with blasting'; Amos 4, 9 'I have smitten you with blasting and mildew'.) Wheat rust was known from Roman times. In the middle of the eighteenth

† A comprehensive early history is contained in Large (1940).

century, cereal diseases were rising to greater economic and scientific prominence in Europe. For example, in 1750 the Academy of Arts and Sciences in Bordeaux offered a prize for the best dissertation on the cause and control of the blackening of wheat ears with bunt. This prompted much work, the most notable of which was a comprehensive set of field experiments by Tillet, after whom the disease *Tilletia caries* was named. Tillet showed in a seminal series of field plot experiments that where seed was dusted with black dust from bunt balls produced in the ears of a previous crop, the new crop would also be infected. Where clean seed from a clean parent crop was planted, the new crop was relatively unaffected by bunt. He also showed how the application of manure from horses fed with bunted grain could increase the incidence of bunt in a crop.

His pioneering work also demonstrated the first fungicidal seed treatment, showing that plots of seed which had been treated with lime, lime and salt, lime and nitre or putrified urine all gave rise to crops which were relatively free from bunt.

This discovery was soon to become of major economic and social importance in rural France.

In 1760 25–50% of the French wheat crop was lost due to bunt (La Carie). Much of the remaining crop was tainted with stinking smut (*Ustilago nuda*). There was therefore a great practical incentive for Tillet's remedy of steeping grain in lime and salt to rapidly become standard practice in a process which was called 'le chaulage'.

In 1816 another fungal disease struck the French cereal crops. Ergot (*Claviceps purpurea*) infested rye and, when eaten, caused gangrene of the extremities of the body and abortions in women amongst the peasant communities of Burgundy and Lorraine. Hunger had forced them to eat bread made from the blackened and elongated kernels of bad rye, there being insufficient good quality grain to buy.

From the middle of the nineteenth century, fungal diseases began to invade the major crops of the developed world to a very serious extent and had a major economic impact on the agricultural industry and on the fabric of society. The catalogue of events during the latter half of the nineteenth century is dramatic.

In 1845 potato blight (*Phytophthora infestans*), first referred to as the potato murrain, made its debut in Europe and swept through Poland, Germany, Belgium, France, England and Ireland. Nothing was known of the cause or the course the disease would take. Potatoes were rotting in the ground and the disease was spreading among potatoes like cholera among man. There was even fear of a link between the two diseases. There were threats of famine in Poland and real famine in Ireland, where potatoes were the staple diet, together with considerably increased hardship amongst the poor elsewhere for whom potatoes were also a major part of their food supply. There were also immense political repercussions as a result of the Irish famine; a famine exacerbated by the poverty of much of the population and their inability to buy food where their own crops had failed. The plight of the poor country folk was also made worse by the pressures of landowners claiming their rents while there was still a chance of extracting money. Vast numbers of people became destitute.

The effects of the spread of potato blight also had repercussions in the British Parliament on agricultural trade legislation. The shortage of potatoes and the artificially high price of cereals due to the protectionist Corn Laws aided the cause of the Anti-Corn Law League in Britain. They gave Sir Robert Peel the opportunity to

(a)

(b)

Plate 1 — The effect of potato blight on crop foliage. (a) Healthy potato foilage. (b) Foliage damaged by potato blight.

renege on his election policy and repeal the protectionist 1815 Corn Laws, removing the protection of the high wheat prices enjoyed by the landowning oligarchy and by allowing free imports, reducing the cost of food for the growing urban workforce of Britain. This also opened the way to a free trade policy for both agricultural and manufactured goods and the rapid expansion of commerce. Shipments of maize were imported from the USA to distribute to the Irish poor in return for road building work or a small cash payment, and massive relief operations were set up. In the short term, however, conditions were still grim and vast numbers sought relief through emigration to Canada and the USA in search of a new life.

Between 1845 and 1860 a million people died in Ireland and one and a half million emigrated largely as a result of a fungal disease on potatoes! The same disease had caused a complete change in British trade policy. Had it been known at the time, the course of history could have been changed by a few tonnes of copper sulphate and lime!

The next crop to suffer a series of ravages from fungal attacks was the grapevine, with obvious impact on the wine industry. Grape vine powdery mildew was first observed in 1845 in Margate, England. In 1848 it was reported near Versailles in France and by 1851 it had spread throughout Europe, appearing as a major problem in France, Portugal, Italy, Switzerland and Germany, threatening the grape industry with ruin. By 1852 the best vines of Madeira had been so ravaged that there was a fear that they would have to be abandoned and replaced with orange trees instead.

European botanists quickly came to recognize the vine powdery mildew as a disease caused by a parasitic fungus and in 1851 J. H. Léveillé, a Parisian doctor, published a methodical classification of the powdery mildews and showed them to have a very different growth habit from that of potato blight.

Fortunately, Mr Tuker, the Margate gardener who first observed the grape mildew, had previously tried sulphur against other mildews and now found it to be effective against the grape powdery mildew as well. Professor Duchartre of the Institut Agronomique at Versailles tried Mr Tuker's remedy and also found it to be effective. The major task was then to develop a remedy which could be practically applied to hundreds of thousands of hectares of vineyards. As described in the next section, the problem was overcome and sulphur applications to control powdery mildew saved the grape production and wine industries from economic devastation.

The general view was that both potato blight and vine powdery mildew were indigenous diseases to the New World and had been imported into the Old World either accidentally or in specimens of fungi that had been collected by cryptogamic botanists of the time. They had then found European conditions to be very favourable to their growth and spread into an open and attractive 'ecological niche'.

In the late 1840s yet another powdery mildew (hop powdery mildew) made its debut and in 1850 another disease, *Botrytis bassiana*, spread through the Lombardy region causing the muscardine disease of silk worms, devastating the wealthy Lombardy silk industry.

During the period between 1865 and 1875 the French grape industry was again devastated, this time by the fungus *Phylloxera* which was transmitted by an insect, the root aphid. The symptoms of this latest scourge were extremely distressing. Vines which appeared healthy in May or June stopped growing prematurely, the leaves turned yellow or red and progressively fell off the vines during the summer.

(a)

(b)

Plate 2 — Grape powdery mildew (*Uncinula necator*). (a) On foliage. (b) Causing damage to young berries.

The crop of grapes which had been progressing well did not ripen. In the year following attack, the vines died because they had become devoid of any fine absorptive rootlets on the root system and had developed swollen nodules on the adventitious roots at the stem base. By 1875 the French vine industry was losing £50 m/year (an astronomical sum of money at that time) and a million hectares of vines were affected. Fortunately for the wine industry and consumers a solution to this plague was found. This was to graft French vine scions onto American *Phylloxera*-resistant rootstocks and to plant these new resistant plants into the old vineyards of France. This enabled the traditional grape varieties and wine blends to be restored.

Other parts of the world were also suffering from the ravages of fungal attack. In 1869 the rust fungus was first identified on coffee bushes in Ceylon (Sri Lanka) and it too spread like wildfire. Between 1871 and 1878 coffee yields in the island declined from 565 kg/ha to 251 kg/ha due to the rust disease which defoliated the trees. The economic result was very dramatic; it was the disappearance of the Ceylon coffee industry and its replacement by the Ceylon tea industry!

By 1878 it was the turn of the battered French vine industry to suffer yet again following the appearance of vine downy mildew (*Plasmopara viticola*). By 1882 this disease had become a widespread problem. Fortunately for the industry a chemical remedy was to hand and was published by Professor Pierre Millardet in 1885 in a communication to the Society of Agriculture for La Gironde:Bordeaux mixture made from copper sulphate and lime.

In 1885, yet another disease appeared on grapes in France, black rot (*Guignardia bidwellii*). This was again a disease from the New World which had found its way across the Atlantic to become established in Europe.

The New World also met its share of trouble from crop diseases. During the 1880s the cotton industry of the USA became beset by soil-borne *Fusarium* vascular wilt; a problem which, fortunately, was able to be contained through the selection of cotton varieties which were genetically resistant to the organism.

At the end of the 1800s and over the turn of the century, the world wheat crop became the next target of the fungi, in the form of severe rust (*Puccinia* sp) outbreaks. Examples of major epidemics were in 1889 in Australia, 1892 in Prussia, 1894 in the USA. In the USA and Canada, in 1904 and 1916, wheat farmers had to contend with the worst epidemics ever recorded. Other major outbreaks recurred over the same period in countries as diverse as Denmark, Russia, Argentina, South Africa and India, illustrating the global scale of economic losses.

The economic need for better and more reliable control measures to combat plant diseases has provided the incentive for progressively more effective remedies to be sought, discovered and adopted into widespread use during the twentieth century to avoid the devastating consequences of plant disease epidemics of the nineteenth century. However, the continued intensification of agriculture and the opening up of further 'ecological' niches through control of the old diseases has led to an increase in awareness of, and a need to control, a further plethora of fungal diseases which result in economic loss. The major diseases of the most important world crops listed in Table 4 give some indication of the range of pathogens which now need to be controlled, to a greater or lesser extent, to achieve efficient crop production.

There is economic demand for both crop quantity and crop quality. As the quality

Table 4 — Major diseases of major world crops

Crop	Latin name	English common name	Part affected	Major effect
Wheat	*Erysiphe graminis*	Powdery mildew	Foliage	Yield loss
	Puccinia recondita	Brown rust	Foliage	Yield loss
	Puccinia striiformis	Yellow rust	Foliage	Yield loss
	Puccinia graminis	Black rust	Stem	Yield loss
	Pseudocercosporella herpotrichoides	Eyespot	Stem	Yield loss
	Tilletia caries	Bunt	Ear	Yield loss
	Rhizoctonia solani	Sharp-eyespot	Stem	Yield loss
	Fusarium sp	Ear scab	Stem, foliage, ear	Yield loss
	Septoria tritici	Leaf blotch	Foliage	Yield loss
	Septoria nodorum	Glume blotch	Ear	Yield loss
Barley	*Erysiphe graminis*	Powdery mildew	Foliage	Yield loss
	Puccinia recondita	Brown rust	Foliage	Yield loss
	Pseudocercosporella herpotrichoides	Eyespot	Stem	Yield loss
	Pyrenophora teres	Netblotch	Foliage	Yield loss
	Pyrenophora graminea	Leaf stripe	Foliage	Yield loss
	Rhynchosporium secalis	Rhynchosporium	Foliage	Yield loss
Apples	*Venturia inaequalis*	Scab	Leaves & fruit	Yield & quality
	Podosphaera leucotricha	Mildew	Leaves & fruit	Yield & quality
	Alternaria mali	Rust	Leaves	Yield & quality
Pears	*Venturia pirina*	Scab	Leaves & fruit	Yield & quality
Stone fruit	*Sclerotinia fructigena*	Brown rot	Fruit	Yield & quality
Grapes	*Plasmopara viticola*	Downy mildew	Leaves & fruit	Yield & quality
	Uncinula necator	Powdery mildew	Leaves & fruit	Yield & quality
	Botrytis cinerea	Grey mould	Fruit	Yield & quality
	Guignardia bidwelli	Black rot	Leaves & fruit	Yield & quality
Potatoes	*Alternaria solani*	Early blight	Leaves & tubers	Yield & quality
& tomatoes	*Phytophthora infestans*	Late blight	Leaves & tubers	Yield & quality

of control of diseases which cause blemishes in the crop, for example apple scab (*Venturia ineaqualis*), improves, so does the standard of consumers' expectation. In the late 1800s there were increasing demands for less blemished fruit as the early pest control techniques began to be applied. Today, total freedom from blemishes is a standard expectation in the developed world, placing a very heavy demand on control measures.

INSECT PESTS

Insects are one of the oldest and most sucessful orders of the animal kingdom and they have had nuisance or economic impact on man from earliest times. Some of the most dramatic scourges have been the plagues of locusts, described from earliest biblical times, but throughout history insects have attacked crops and reduced yield or quality. Attempts have also been made from early times to combat them. Pliny in

Plate 3 — Scab damage on apples.

79 A.D. advocated the use of arsenical compounds. They were also widely used by the Chinese in the sixteenth century.

The early concerns about the economic impact of insect pests were probably greatest among fruit and vegetable growers, where insect damage not only reduced yields but seriously reduced the quality and appeal of the crops. As described in the next chapter it was in fruit and vegetable crops that the use of insecticides discovered in the late nineteenth century and early twentieth century first became established.

Changing agronomic practices also contributed to the increase of insect problems. One example is the spread of the Colorado beetle, a native pest of the Rocky Mountains of the western USA. As potato cultivation spread westwards with the early settlers, so the Colorado beetle took advantage of the attractive new habitat provided by the crop to spread eastwards. From its first discovery in the Rockies in 1824 it spread to Iowa and into Canada by 1861 and into Illinois by 1885. By 1877 it had spread across the Atlantic and had been reported in Germany.

Cotton has always played host to a number of foliage- and boll-eating insects and it proved economic to use arsenical insecticides on a significant scale on cotton in the USA in the early twentieth century.

The rapid and widespread use of insecticides to improve crop yield and quality did not take place, however, until the first developments of organic insecticides, notably DDT and other organochlorine compounds, in the 1940s.

The very strong economic pressures caused by the loss of crop production to pests

(a) (b)

Plate 4 — Effect of insect damage on cotton foliage. (a) Defoliated crop due to insect attack. (b) Healthy crop treated with insecticide.

and diseases provided the impetus for technical advance in methods for their control and for the development of a crop protection chemicals industry. The early growth of this industry is the subject of Chapter 3.

3

The advent of crop protection chemistry

THE EVOLUTION OF A TECHNOLOGY

The development of chemical methods of crop protection to combat the economic ravages of pests, weeds and diseases has paralleled and been dependent upon the development of many other branches of biological and chemical science. Without the advance of those sciences, crop protection chemicals as they exist today would not be possible. During the nineteenth century the increasing intensification of agriculture and the development of new pest and disease pressures led to intensified study of cryptogamic botany and entomology as part of a scientific effort to improve agricultural production. At the same time scientific thought was being applied to plant breeding and plant nutrition in an increasingly formal way.

The development of a 'scientific method' and an international scientific community aided the discussion and dissemination of ideas. Journals such as *Annales des Sciences Naturelles*, founded in 1824; *Journal of the Royal Agricultural Society of England*, 1840; *The Gardeners' Chronicle and Agricultural Gazette*, 1841; *Journal of the Royal Horticultural Society*, 1845; *Hedwigia Ein Notizblatt für Kryptogamische Studen*, 1852; *Proceedings of the Royal Society*, 1854 and *Jahrbuch für Wissenschaftliche Botanik*, 1858 started to be published.

The old academic institutions of Europe began taking an interest in the science of crop protection from the mid-1800s. The Royal Agricultural Society of England was formed in 1840 and in the USA, following the Civil War, the United States Department of Agriculture was founded (in 1862) with a programme to set up state agricultural experimental stations with specialist entomologists and plant pathologists.

The rate of scientific advance before about 1850 was almost imperceptible. During the century between 1850 and about 1950 the early foundations of crop protection chemistry were being laid. During the quarter century between 1950 and 1975 there was an explosive development of a sophisticated crop protection industry encompassing a complex set of inter-relationships between chemical companies, governments, universities, advisory organizations and farmers. Each group has played a different role. Chemical companies have applied their inventive skills to producing new molecules with valuable agronomic properties, developing recommendations for their use, demonstrating benefits, providing support to farmers, and generating supporting data on the safety of products to consumers and the environment. Government has played a part in helping to define the agronomic needs and benefits, advising on product use and in constructing and policing policies to protect the consumer and environment from any adverse side-effects. Universities and research institutes have contributed to the advances in biological and chemical science which have underwritten the industry and provided practical empirical field testing of compounds and advice to farmers.

THE EARLY YEARS

Before 1850 a few advances had been made in the science of crop protection. In 1667 Robert Hooke first viewed fungal symptoms under the newly invented microscope and described 'several kinds of small and variously figured mushrooms'. He did not, however, recognize the nature of the organisms he saw but rather thought the symptoms were outgrowths produced by the vegetation itself. A century later in 1767 an Italian, Felice Fontana, examined wheat rust under the microscope and noted 'minute vegetables with bodies resembling seeds'. In 1807 in France, Bénedict Prévost first observed, under the microscope, fungal spores germinating in water. By the mid-1800s wheat rust and bunt had begun to be recognized by European botanists as being fungal diseases. In 1845 Dr Montagne in Paris first described the detailed morphology of the potato blight fungus.

In 1752 Tillet had made his remarkable series of experiments to display the cause of wheat bunt and had demonstrated the first practical application of a seed treatment (lime and salt) for its control. Prévost, through good practical observation, discovered an improvement over the basic formula. He had noted the particular success of a steep treatment carried out in an old copper pannier and wondered whether the copper enhanced the efficacy of the treatment. Experiments adding polished copper and copper covered with verdigris to glass beakers of water containing bunt spores showed marked fungitoxic effects, with greater potency using copper covered with verdigris. Prévost then tried to make copper acetate by dissolving verdigris in vinegar, which he found to be even more effective, even down to concentrations as low as 1 part copper to 4 million parts solution (0.25 ppm). From this point onwards copper came to play a major role in crop protection. Copper acetate and then, more widely, copper sulphate, because of its low cost, became used as a seed soak. Because of the risk of phytotoxicity the seed soaking time had to be carefully controlled and the development of acidity had to be neutralized with lime.

Later, Dombasle in France produced a safer seed treatment made from sodium sulphate and lime.

It was not until after 1850, however, that chemicals started to play an important part in crop protection.

THE 'FOUNDATION CENTURY' FOR CROP PROTECTION 1850–1950

The period from 1850 to 1950 saw many of the major advances which established chemical crop protection as an integral part of agricultural production. It saw the

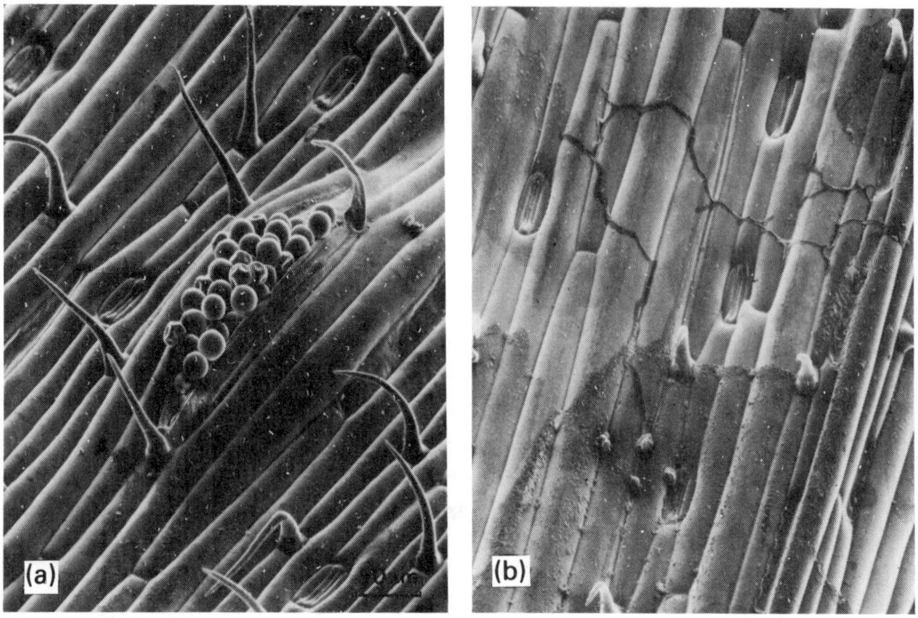

Plate 5 — Scanning electron micrograph showing (a) brown rust (*Puccinia recondita*) spores bursting through the surface of a leaf from internal infection, (b) rust spores which have landed on a leaf and germinated but have been prevented by a fungicide from entering and infecting the leaf tissue.

discovery of most of the simple inorganic crop protection chemicals and the major transition from inorganic to complex organic chemicals. It also saw the establishment, on the basis of early, simple successes, of the commercial organizations whose research and development strength would become the engine of growth of the industry after 1950, a topic discussed in greater depth in Chapter 9.

Control of fungal diseases

This period opened with some of the first major works on mycology. In 1853 Anton de Bary published *Die Brandpilze*, Rev M. J. Berkley *Introduction to Cryptogamic Botany* in 1857, and Julius Kühn *The Diseases of Cultivated Plants, Their Cause and Prevention*.

In 1854 Ferdinand Cohn observed for the first time a fungal germ tube invading a plant cell: *Pythium* attacking the cell of a fresh water alga. In 1860 Anton de Bary observed the life cycle of the potato blight fungus and by 1865 had traced the life cycle of wheat rust (*Puccinia graminis*).

The first large scale development of a fungicidal spray came from the observation of Tuker in 1845 and Duchartre in 1848 that sulphur could effectively control vine powdery mildew. The challenge was to produce a product which could be practically applied to thousands of hectares of vineyards to combat the spreading powdery mildew. Many approaches were tried including blowing sulphur dust with special bellows. The sulphur did not stick to the plants and was easily washed off by rain. However, it was found that better adhesion could be achieved if the crop was dusted when the foliage was still wet with dew. Even then, the dust was washed off and could not protect the new succulent growth which appeared after sulphur application. An alternative approach was found to be to make lime-sulphur by boiling sulphur and lime in water and applying the resulting liquid with a syringe. By 1855 Bequerel had found that boiling sulphur with an alkali, potash or lime, and then precipitating it with acid produced a very fine form of sulphur which got into every crevice of the plant wetted by the spray, thereby giving better control.

Despite these improvements, however, simple dusting with fine dry sulphur became accepted as the easiest practice. Over the years a traditional routine dusting programme grew up in France using three applications per year, one in the spring when the young shoots were 5 to 10 cm long, the second at about flowering time and the third at colour change when the grapes began to mature. This then became further modified over time and between countries to suit local disease epidemiology and increasing standards of disease control expectation.

The next major development in fungicide sprays came in 1885 with the establishment of Bordeaux mixture (copper sulphate and lime) for the control of vine downy mildew (*Plasmopora viticola*). From its first occurrence in France in 1878, downy mildew had become an established disease. Pierre Milardet, Professor of Botany at Bordeaux (and an ex-student of Anton de Bary) was studying the life cycle of the disease and seeking a cure. When visiting a vineyard of Saint-Julien in Médoc he noticed that the vines next to the path were healthy while those in the rest of the vineyard were suffering severe *Plasmopora* symptoms. These healthy plants had a blueish white deposit of copper sulphate and lime on them, sprayed to deter pilferers. He immediately experimented with calcium, iron and copper salts alone and in mixture with lime as treatments for downy mildew and concluded that copper sulphate plus lime gave the best results. He therefore recommended it as Bordeaux mixture and its use became standard practice in vineyards. It also opened the way to considerable efforts to try to make improved versions. Little of this effort however, achieved any significant advance and the standard mixture continued in use for a century.

The success of Bordeaux mixture in France led to it, and variants, being tried worldwide for control of fungal diseases on a wide range of crops. In 1885 Millardet himself demonstrated the effectiveness of copper sprays against potato blight and in 1888 and 1889 Aimé Girand and Prillieux in France conducted large scale experiments with the co-operation of farmers to demonstrate the yield benefits to be gained from potato spraying.

In 1886 a Section of Vegetable Pathology was set up as part of the Division of Botany in the United States Department of Agriculture. Very strong links were forged between US and French workers in the field and the US workers organized a large programme of trials with the co-operation of leading farmers, using all the variants of the French fungicides through the best of their application equipment.

The US trials programmes systematically explored dose responses to minimize the cost of spray required per hectare consistent with efficacy and safety from phytotoxicity. Research workers also investigated optimum timing of application in relation to the life cycle of the particular fungus concerned and drew up spray programmes accordingly. One of their major targets was control of black rot (*Guignardia bidwelli*) on vines, a disease indigenous to North America. Other targets were apple and pear scab, pea leaf blight, gooseberry mildew and leaf blight of cherry and plum.

Despite considerable effort in the late 1800s, attempts to control the major problem of wheat rust by sprays were not successful. This was partly because sprays could not be persuaded to stick to the leaf surfaces even with the addition of soaps but, more importantly, because an effective active ingredient had yet to be discovered. Until the latter half of the twentieth century, wheat rust could only be contained by a combination of early planting and harvesting and a measure of genetic resistance.

In the USA the stage was set for an expansion in Federal-Government-sponsored crop protection research by the 1888 Hatch Act, which set up State Agricultural Experimental Stations in conjunction with the Land Grant Colleges in every state.

In the UK, Government support for crop protection research was formalized in 1889 by the setting up of the Board of Agriculture.

As far as fungicides were concerned, such official support led to greater use of sulphur and copper as standard agricultural practice and these two elements, in simple organic forms, grew to dominate fungicide use for a century.

The first advances of 'organic chemistry' into the field of fungicides came in new seed treatments for control of wheat bunt; the disease first controlled in 1752 with copper compounds. Ehrlich in Germany had worked with compounds made from arsenic and various dyestuff intermediates in a search for a pharmaceutical product that would control syphilis in humans without being too toxic to man. This led to the invention of 'Salvarsan' in 1909. German plant pathologists followed similar chemistry, testing compounds made with metals known to have fungicidal properties combined with benzene or with organic dyestuff intermediates. In 1912 Dr Rielm of the German agricultural research service tried many new compounds to no effect. However, one, a compound of mercury, chlorine and phenol submitted by the Bayer company, proved, when treated with a little alkali to solubilize it, to be an effective seed steep for wheat bunt control at 1 ppm in water. This discovery led to substantial development of organomercury seed treatments. The first organomercury seed treatment was launched in 1915 by Bayer under the trade name 'Uspulam', to be followed in 1924 by another under the name 'Tillantin R', and in 1929 by 'Ceresan' containing a nitrophenol mercury compound, PMA. In 1933 ICI introduced 'Agrosan G', containing a mercury compound with unsubstituted rather than substituted phenol. These compounds continued to dominate the cereal and soil-borne disease control markets until mercurial compounds were banned (on grounds of user hazard) by various governments in the 1970s and 1980s.

Insecticidal control

Despite the fact that insecticidal substances had been known from antiquity, it was not until the 1880s that insecticides started to make a significant impact in agriculture, initially on horticultural crops.

The earliest insecticides were either natural products extracted from plants or very simple inorganic compounds. In about 2500 B.C. the Sumarians used sulphur to control insects and mites. In *c* 1500 B.C. the Chinese developed insecticides derived from plants for protecting plant seeds and the fumigating of plants infested with insects.

The insecticidal properties of dried flower extracts from *Chrysanthemum cinerarifolium* had been known for centuries and powder from dried flowers was produced commercially in Persia before 1800. First imports into Europe and the USA began in the mid-1800s.

Use of derris dust, made from the root of *Derris elliptica* and certain other derris species, was reported in the seventeenth century and was known in China considerably earlier. The active ingredient, rotenone, was isolated in 1895 and its structure established in 1933. The insecticidal properties of water in which tobacco leaves had been soaked were known in the middle of the eighteenth century; the active element was identified as nicotine in 1828.

The use of arsenical compounds had been reported in China from A.D. 900. The use of arsenical products in Europe and the USA began in the early nineteenth century. Paris green, an arsenical pigment cupric acetoarsenite, was shown to be effective in controlling an outbreak of Colorado beetle in 1867. In 1878 the London firm of Hemmingway & Co sent the arsenical residue of their magenta dye production for test and it too was found to be useful as an insecticide, and became referred to as London purple. Lead arsenate was used in the USA in 1892 against the gypsy moth plague.

In 1855, peach growers in California were using a sulphur/lime mixture originally developed as a sheep dip, for the control of scale insects and peach leaf curl.

During the 1880s fruit growers in the United States started using spray oils made from kerosene, soap and water or a kerosene emulsion made with sour milk. Casein in the milk was a better emulsifying agent than soap. The oil films formed killed insects by suffocation and by dissolving some of their protective coatings. Because the oil sprays were phytotoxic to the crop, they could be used only during the winter period, between harvest and the next season's bud break, giving control of woolly aphis, scales, mosses and lichen. The result, however, was a new ecological environment for the summer season which left an ecological niche for other pests and diseases such as codling moth (*Lydia pomonella*), red spider (*Panonychus ulmi*) and the fungal disease apple scab (*Venturia inaequalis*). Paris green, first used against Colorado Beetle, had been shown in 1879 to be effective against codling moth larvae. By 1892 spraying fruit trees before the apples were pea size with Paris green, white arsenic or London purple had become standard practice in the USA. These sprays were first employed in England in 1895 and were widely used in all commercial orchards by 1910.

The copper in Paris green, noted in the previous section as an effective fungicide, also checked the growth of the apple scab fungus and led to a growing expectation for relatively scab free apples among urban consumers in an increasingly industrial society. This in turn encouraged pathologists to study apple scab control.

The widespread use of insecticides on a broad range of crops is, however, a phenomenon of second half of the twentieth century.

Chemists first started to define the molecular structure of natural insecticides in the early twentieth century. The structure of nicotine was established in 1904 by Pictet and Rotschy.

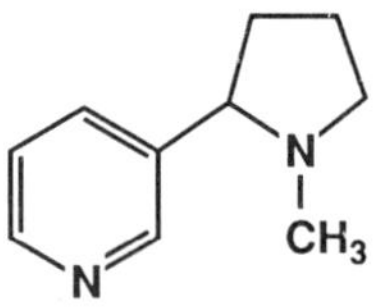

Nicotine

The active elements of pyrethrum were defined in 1924 to be esters of two acids, chrysanthemic acid and pyrethric acid, and two alcohols, pyrethrolone and cinerolone.

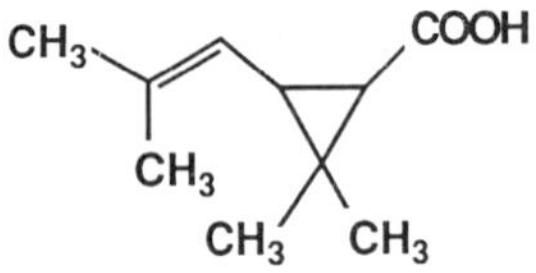

Chrysanthemic acid

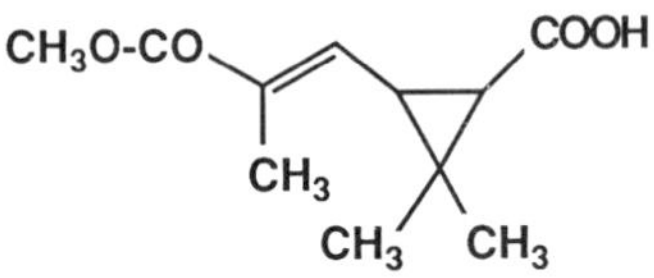

Pyrethric acid

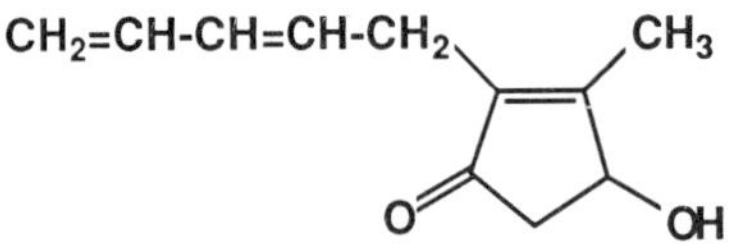

Pyrethrolone

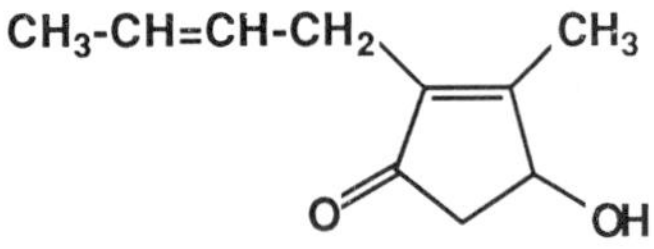

Cinerolone

The first breakthroughs in synthetic insecticides came in 1939 with the discovery of the high insecticidal activity of organochlorine compounds. As in the case of herbicides, covered later, the major incentive for invention was the pressure of needs arising during the Second World War. The search for good insecticides was stimulated when supplies of derris were interrupted.

The most famous organochlorine insecticide is almost certainly DDT [dichlordiphenyltrichlorethane now given the systematic name 1,1,1-trichloro-2,2-bis (*p*-chlorophenyl) ethane].

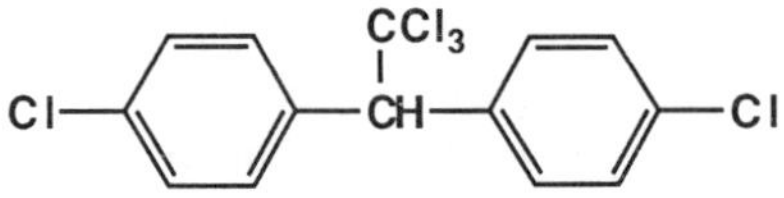

DDT

The compound was first synthesized in 1873 by a German postgraduate student who had no idea of its insecticidal properties. In 1939 a Swiss entomologist, Paul Müller, rediscovered the compound in a search for a long-lasting insecticide to control clothes moths. DDT subsequently proved to be effective against a wide range of insect pests, notably flies and malaria-carrying mosquitoes. This won Dr Müller the Nobel Prize for Medicine in 1948 for his lifesaving discovery.

DDT was cheap to manufacture and extremely effective on a very wide range of insects. It became widely used in public health, controlling mosquitoes, which transmit malaria and yellow fever; body lice, which carry typhus; and fleas, which are vectors of plague. It also became a standard chemical for agricultural use against caterpillars on cotton, vegetables and fruit; against Colorado Beetle on potatoes; and gypsy moth and spruce budworm in forests. By its year of peak usage in about 1960, there was an annual production of over 7000 tonnes.

A number of organochlorine compounds related to DDT were also found to have valuable insecticidal properties and made a significant contribution to agriculture before being superceded by other, less environmentally persistent classes of chemical. Neither DDT nor its major metabolite are readily decomposed by micro-organisms, heat and light, or by metabolism in plants or animals. The compounds can accumulate in body fat and be passed from prey to predators in the food chain.

The other major group of insecticides to be invented in the 1940s was that based on organophosphorus chemistry (Ware 1983). The first observations of the insecticidal properties of organic derivatives of phosphoric acid were made in Germany during World War II when materials related to nerve gases were being tested in a search for substitutes for nicotine, which was in critically short supply.

World War II
German nerve gases

ortho phosphoric acid

$$HO-\underset{\underset{OH}{|}}{\overset{\overset{O}{\|}}{P}}-OH$$

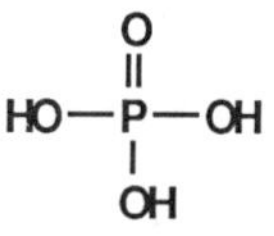

Sarin

$$(CH_3)_2CH-O\underset{CH_3}{\diagdown}\overset{O}{\underset{}{\overset{\|}{P}}}-F$$

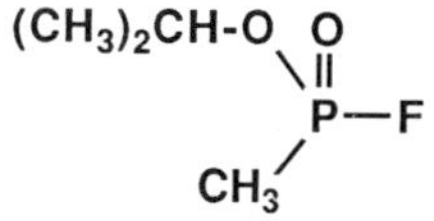

Soman

$$(CH_3)_3C-\underset{}{\overset{\overset{CH_3}{|}}{CH}}-O\underset{CH_3}{\diagdown}\overset{}{\underset{}{P}}-F$$

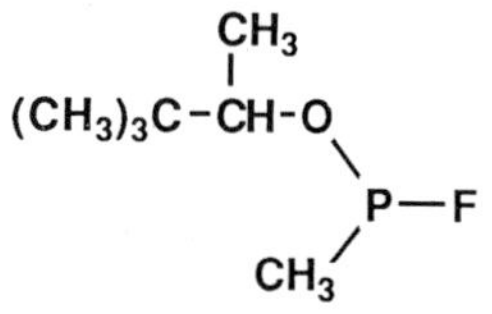

Tabun

$$\underset{(CH_3)_2N}{\overset{C_2H_5O}{\diagdown}}\overset{O}{\underset{}{\overset{\|}{P}}}-CN$$

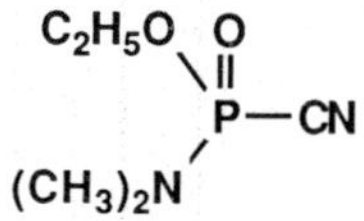

The organophosphates act by inhibiting certain important enzymes in the nervous system, the cholinesterases. The range of organophosphate chemical structures developed is very varied. The first organophosphate compound to be introduced into agriculture was TEPP (tetraethyl pyrophosphate) in 1946.

Pyrophosphate

$$O-\underset{\underset{O}{|}}{\overset{\overset{O}{\|}}{P}}-O-\underset{\underset{O}{|}}{\overset{\overset{O}{\|}}{P}}-O$$

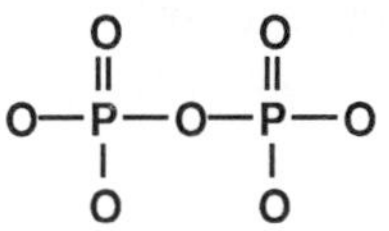

TEPP

$$(C_2H_5O)_2\overset{\overset{O}{\|}}{P}-O-\overset{\overset{O}{\|}}{P}(OC_2H_5)_2$$

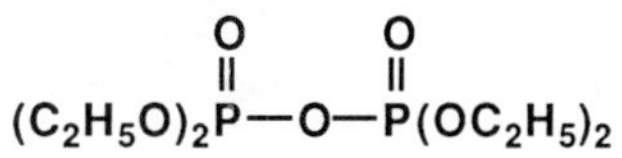

The second introduction, ethyl parathion (1946), and its close but safer and broader spectum analogue methyl parathion (1949), were to become the widest used organophosphate insecticides and would dominate the world insecticide market of the 1950s and 1960s.

Malathion, introduced in 1950, was quickly adopted for use on a wide range of fruit and vegetable crops. It was also safe enough for use in stored grain and for use on pets and humans to control fleas and lice.

These inventions proved to be the foundation for the growth of insect control chemistry over the next quarter century. Many of the compounds invented also showed systemic mobility within plants, enabling them to control pests not in direct contact with spray, for example larvae living inside plant stems or leaves.

Weed control

Herbicides began to make their appearance in the first half of the twentieth century in a way intimately bound up with the search for 'plant growth substances'. In 1897 it had been discovered in France that a 2% solution of copper sulphate would kill charlock (*Sinapis alba*) in wheat without damage to the wheat, partly because the spray solution stayed on the horizontal broad leaves of the charlock but ran off the vertical narrow leaves of the wheat. Selective weed control with copper sulphate did not, however, become important. The real breakthrough came with the introduction of the 'hormone' weedkillers in the 1930s and 1940s; which were a spin-off from research into the mechanisms influencing plant growth. As with the study of plant pathology, the study of plant growth had made some limited advances before the mid-1880s, but the century from 1850 to 1950 was a period of major progress. In 1755 Francis Home was requested by the Edinburgh Society for Improvements of Arts and Manufacturing 'to try how far chymists will go in settling the principles of agriculture'. He sought to understand plant nutrition, to analyse plants for their major chemical components and to test various substances for their nutritional value, work which was also progressed by others of the time such as Wallerius in Sweden and Priestly in England. Progress was slow, however, because knowledge of chemistry was insufficiently advanced.

By 1804 chemistry had reached the point at which Theodore de Saussure of Geneva was able to prove that plants derive carbon and oxygen from the air and minerals and nitrogen from the soil. Davy in the UK and Thaer in Germany collected information on practical manuring and put that into chemical terms. Boussengault in France did field experiments to relate the chemical content of manure to that of the plant.

In 1840 Liebig suggested adding the minerals required by plants to the soil as soluble salts. J. B. Lawes, on his Rothamsted estate in England, explored the manurial benefits of a range of alternative substances including bones, shown by chemists to contain calcium and phosphorus. It was demonstrated that treating bones with sulphuric or hydrochloric acid would yield soluble phosphates which were then available to promote plant growth. Once geologists had found vast deposits of mineral calcium phosphates, Lawes patented (in 1842) 'superphosphate' made from mineral phosphate and sulphuric acid, set up a factory to make in and proceeded to sell it.

Lawes also showed that sulphate of ammonia, a cheap by-product from coal-gas production, was a valuable fertilizer because of its available nitrogen content. The yield increases achievable through application of fertilizer were immense. In some experiments wheat yields were raised by 50%, from 20 to 30 bushels/acre (175 to 260 m^3/ha). Fertilizer use became a standard part of the expansion of agricultural production in the late nineteenth century. Chemical manufacture grew and Chilean nitrate and guano industries thrived.

By the middle of the ninteenth century, scientists were beginning to delve further into the chemical nature of plant growth. The German botanist Julius Sachs began to study growth movements of plants and hypothesized an 'organ-forming substance', the action of which controlled organ development and which was influenced by external factors such as gravity. Charles Darwin noted that plant shoot growth responded to strong light (phototropism) and that, although bending occurred below the shoot tip, removal of the shoot apex would prevent the phototropic effect. He published the hypothesis that a chemical messenger was involved.

In 1919 the influence of the shoot apex on growth was shown more clearly and in 1926 the Dutch scientist F. W. Went isolated 'the growth substance' by letting it diffuse down from shoot tips into agar blocks. Chemical science was not sufficiently advanced, however, to identify the substance.

During the 1930s European botanists consolidated their ideas on the nature of growth hormones and the role of chemical messengers. Work started on biochemical studies of hormones, their source, action and fate. This led to a race to find and exploit substances which would enhance plant growth. These substances were given the term 'auxins' from the Greek 'auxein', to grow.

The first effective growth substance isolated from urine was called 'auxin A' and the second, 'auxin B', was isolated from maize germ oil and barley malt. The third substance to be identified, heteroauxin, was indol-3-yl acetic acid (IAA). This was to find its first practical use as a rooting stimulant for cuttings.

Plant physiologists became increasingly fascinated by the new ideas that plant growth could be modified by controlling biochemical processes within the target plants, and the young but growing chemical industry also started to take an interest, especially those companies which were already developing significant business providing major nutrients as 'artificial fertilizers'.

ICI in the UK, for example, started to test large numbers of possible substances as potential growth regulators. The early work, however, produced only IAA and the related NAA, 1-naphthyl acetic acid, which showed effects as a retardant.

World War II was to be a major stimulant to research on chemical control of crops with two objectives: to stimulate crop production in the face of rationing and to find ways of destroying the crops of the enemy.

In the UK in 1940, ICI experiments attempted to kill wheat and oats with NAA at 1–20 lb/acre (1–22 kg/ha). The only effect noted was distortion of creeping buttercups, which recovered again in a few weeks. This led on, however, to experiments to explore the selective weed control properties of NAA. These showed growth effects and phytotoxicity on broad-leaved plants. Rates of application of 10–15 kg active ingredient (ai) per hectare were too high, however, for NAA to become a practical herbicide.

At the same time, at Rothamsted in the UK, IAA and NAA at less than 0.1 ppm were shown to be very toxic to seed germination and, again, the idea arose of using them as herbicides. Studies in unsterilized soil, however, did not bear out the promise of agar and solution culture experiments. This was attributed to excessively rapid degradation in soil and it was postulated that chlorinated analogues might be more resistant to degradation, a hypothesis soon borne out by tests which showed the effectiveness of 2,4-dichlorophenoxyacetic acid (2,4-D) and MCPA.

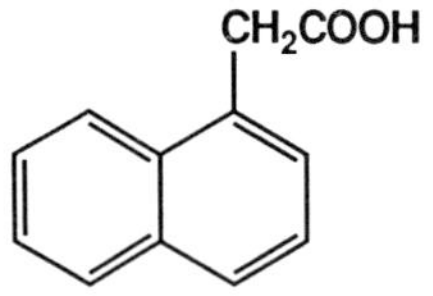

NAA
(Active at 10-25 lb ai/acre. Too costly for practical use)

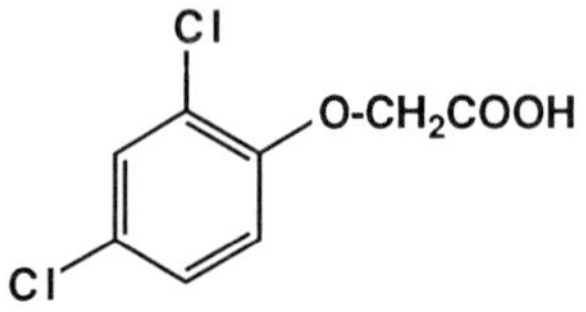

2,4-D

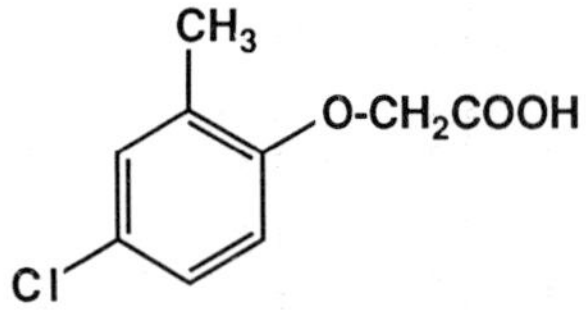

MCPA

The efforts of ICI and Rothamsted were then co-ordinated through the Agricultural Research Council under tight wartime security with the objectives initially of developing a crop destruction weapon against sugar beet (a plan which eventually foundered as being impracticable) and later, as a crop production aid for selective weed control in cereals.

Following a programme of field trials in 1944 and 1945, MCPA was launched in the UK in 1946 as the first 'scientifically produced' selective herbicide.

Meanwhile, a similar train of events had been taking place in the USA at the Boyce Thompson Institute, the University of Chicago, the United States Department of Agriculture at Beltsville, Du Pont and Am Chem. As in the UK, the US National Academy of Sciences War Research Committee was particularly interested in the potential use of plant hormones as crop destruction weapons to support General Douglas MacArthur's South-East Asian island-hopping campaign. They were particularly interested in the toxicity of 2,4-D and 2,4,5-T (2,4,5-trichlorophenoxy acetic acid) to rice, a foreshadowing of the use of 'agent orange' in the Vietnam War of the 1960s.

As in the UK, the selective weed killing properties of 2,4-D became clearer during 1943 and 1944 and the US chemical companies were beginning to recognize the opportunity. In March 1944 Am Chem filed a patent on methods and compositions for killing weeds covering halogenated phenoxy monocarboxylic aliphatic acids, esters and salts. In 1945 they marketed 2,4-D as 'Weedone'. The first 'scientifically developed' herbicides were now launched in both the UK and USA.

After 1945 development work in the UK concentrated on MCPA, using cresol, a by-product of the coal industry as a chemical feedstock. Work in the USA concentrated on 2,4-D, using as its feedstock cheap phenol, a by-product of the oil refining industry.

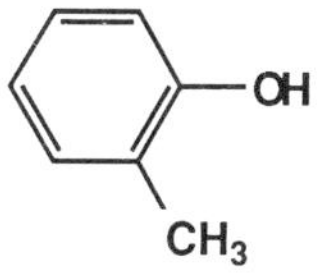

Cresol

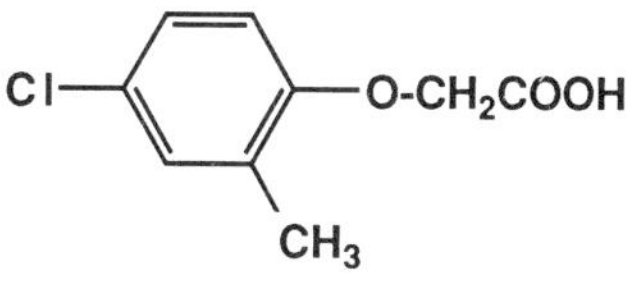

MCPA

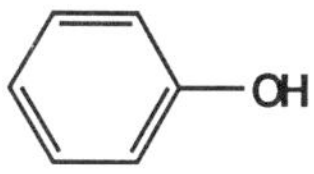

Phenol

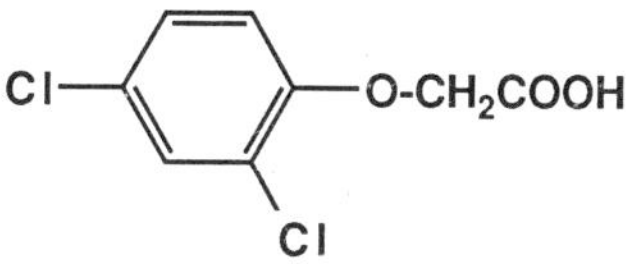

2,4-D

The phenoxy acid compounds have now grown to be the world's major herbicides controlling weeds in cereal crops in most countries of the world.

Biochemical knowledge then aided the invention of homologues which were safe to use in legume crops. The early twentieth century German chemist Knoop had demonstrated in animal systems the process of β-oxidation, whereby enzymes split off two carbon units from an alkyl chain. Zimmermann and Synerholm in the USA in 1947 showed that homologues of 2,4-dichlorophenoxy alkanoic acids with even numbers of carbon atoms in the alkyl side chain caused greater distortion (epinasty) on tomatoes than those with odd numbers of carbons. Wain at Wye explored the link between these facts and showed that even-number carbon homologues of 2,4-D were broken back to the herbicidal acetic acid while odd number chains were broken down to inactive phenol. Joint work with May & Baker then showed that, although legumes are sensitive to the acetic homologue, they lack the enzyme system which β-oxidizes higher homologues. 2,4,5-TB (2,4,5-trichlorophenoxyl butyric acid) was invented as a 'legume safe' hormone weedkiller.

1950 ONWARDS

By 1950 it had become established that organic chemistry could potentially provide a very wide range of agriculturally valuable 'effects chemicals' that could control the major fungus, insect and weed pests that constrained yields, and might also offer, in the longer term, scope for manipulating the growth of crop plants themselves.

During the next 25 to 30 years the number of new pesticide products developed and marketed increased exponentially and the proportion of the crop area treated with chemicals moved from a negligible proportion in most crops (with the exception of vine fungicides, fruit insecticides and cereal seed treatments) to almost total coverage with at least one application for some purpose in most countries of Europe and in North America.

For example the growth in the proportion of the area of major crops treated with herbicides was illustrated in Figs 10 (page 25) and 11 (page 26). An indication of the increase in numbers of chemicals available was given in Fig. 9 (page 24).

Most of the products introduced up to the mid-1950s were solving problems which had not been solved by chemists before. They were creating totally new markets and tapping latent demand. Subsequent inventions have both filled new niches and provided improvements over the existing product range, either expanding the market or capturing a share at the expense of an existing product, or a combination of the two.

The approaches taken by chemical companies, who were now establishing strong R & D organizations, to the active discovery process is the subject of the next chapter. Broadly they have been seeking products which have some form of functional improvement along the lines of the categories listed in Table 5 (page 58). The most compelling area of improvement has been in function performance. This can be illustrated by some herbicide, fungicide and insecticide examples.

Pesticide product improvement

The companies introducing the earliest products were entering a virgin market with strong latent demand arising from pressures to increase yields and to reduce labour inputs in view of rising labour costs per unit of output. This is illustrated in Fig. 14 by the index numbers of agricultural prices and wage rates for the USA. They were able to obtain monopoly profits and to achieve rapid market growth unhampered by strong competitive pressures.

Subsequently introductions of new products could succeed only if they could offer a degree of advantage in order to claim a share of the market. The categories of 'improvement' which have occurred are listed in Table 5.

There is considerable variation in the relative importance of each of these categories as between the different classes of biological effects. For example, improvements of quality and duration of control have been a major factor in fungicides. Improvements in spectrum of control and flexibility of application timing have been a prime force behind herbicide developments, while insecticide research has been seeking replacements for products to which pests have become resistant or which have been environmentally unacceptable. The following sections illustrate some of these improvements, taking examples from each biological effect. Total coverage would require a book in its own right.

(a) Improvement in function performance: some herbicide examples

In row and tree crops, herbicides are essentially a labour-saving technology replacing some form of mechanical weed control. It is only in small grain cereals (wheat, barley and oats) which are grown at too high a planting density to permit practical mechanical methods, that herbicides can be considered a yield-promoting technology†. It is interesting to note the marked impact that herbicide use, combined with

† Rice falls in an intermediate category where hand and mechanical weed control is practised, aided by flood water management.

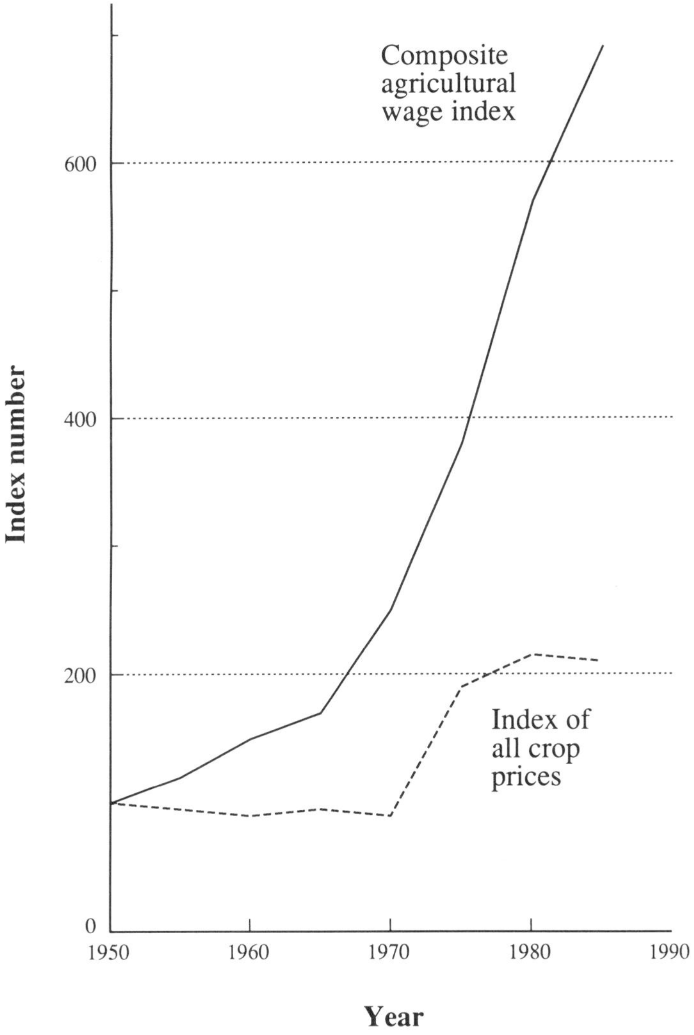

Fig. 14 — Relative growth rates of agricultural product prices and wage rates in the USA.

the greater responses that this permits to the use of higher doses of fertilizer, has had on wheat and barley yields as exemplified by the UK yield data, Fig. 15.

Estimates of the economic value of the increases in yield which could be expected from weed control, at constant fertilizer application rates, made in the early commercial years of MCPA and 2,4-D application to cereals in the UK, suggested a yield benefit of about 20%, an increase in gross return of £12–15/ha at 1950 yields and prices, for an expenditure of £3.7/ha on chemical and application cost.

The development of effective weed control was also a necessary adjunct to the increasing use of combine harvesters (a major labour-saving technology) (Fig. 16) which cannot function effectively if they have to contend with a high density of green

Table 5 — Categories of pesticide improvement

Technological dimension	Type of product improvement
Function performance	Quality of control
	Spectrum of pests controlled
	Duration of control
	Control of pests resistant to previously available treatments
	Safety to crop
Cost	Lower cost per hectare per treatment by
	— Cheaper compound/kg
	— Lower usage rate/ha
	— Fewer treatments per season
Ease of use	Ease of use of formulation
	Flexibility of application timing
Operating cost	Cost of application
Reliability	Independence of soil type, soil moisture, climatic conditions and crop variety
Serviceability	Environmental acceptability/safety
	Safety to users
Systems compatibility	Compatibility with other methods of pest control, including safety to beneficial and predator insects
	Compatibility of spray timing with timing of other cropping operations

weedy vegetation which clogs the threshing mechanism, both impeding physical progress and reducing the efficiency of grain recovery from the threshed crop.

Selective weed control in cereals predated that in other crops because of the nature and timing of the first significant herbicide discovery — that of the phenoxy-acid herbicides 2,4-D and MCPA in the 1940s. This group of herbicides shows a very marked selectivity, effectively removing any dicotyledonous (broad-leaved) weed from a monocotyledonous crop, cereals. The first truly effective herbicides for broad-leaved crops appeared around 1960. The most important, and their dates of introduction, are listed in Table 6 (page 58).

Herbicide use in sophisticated agriculture has become standard practice for these crops, having progressed from initiation to almost total adoption in the USA and UK between 1960 and the mid-1970s (Figs 10 and 11).

The earliest effective selective herbicides were good, but not perfect. Improvement in functional performance has been possible through improvement in the range of weeds controlled (the 'spectrum' of the compound), the margins of safety to the crop (the difference between the rate of chemical application required to control the target range of weeds and that which causes some phytotoxicity to the crop) and the duration of control.

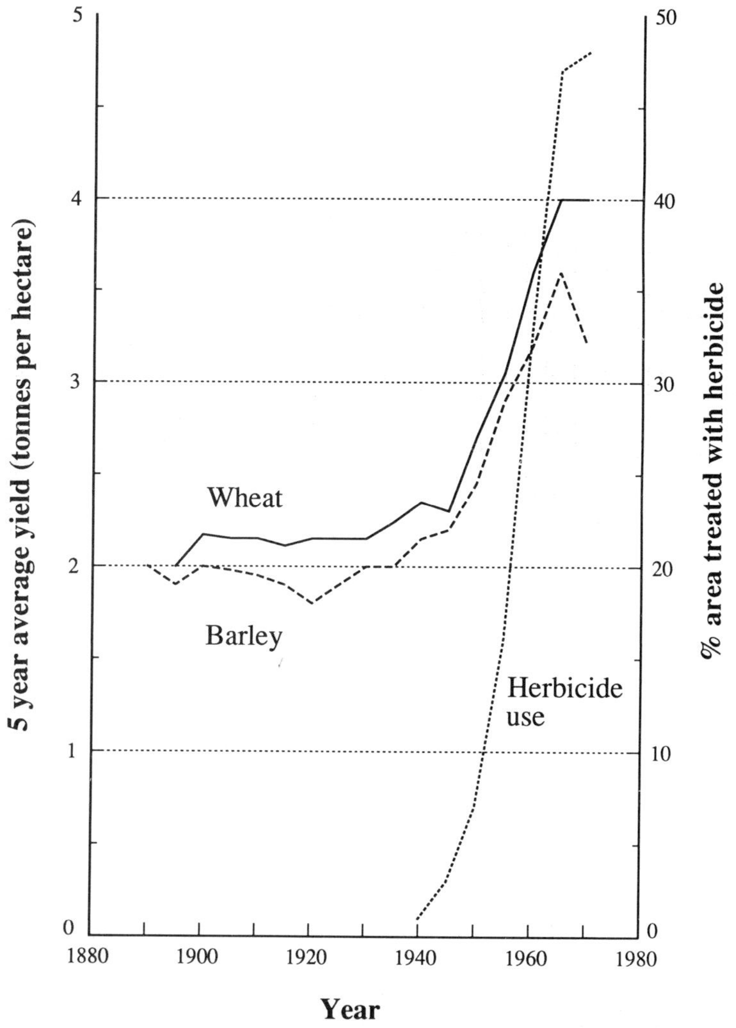

Fig. 15 — The coincidence of advances in UK cereal yields and adoption of herbicide use.

The importance of breadth of weed control spectrum in sustaining the introduction of new herbicides is exemplified by reference to three important crops, cereals in the UK, and maize and soyabeans in the USA.

(i) Cereals in the UK: changes in weed spectrum

The spectrum of weeds occurring in any one crop situation is the result of a balance between the factors favouring and inhibiting the propagation and growth of each species in that complex. The relative strengths of different factors can be significantly modified by agricultural practice. The evolution of cereal weed flora with the changes in cereal farm husbandry practices provides a good illustration.

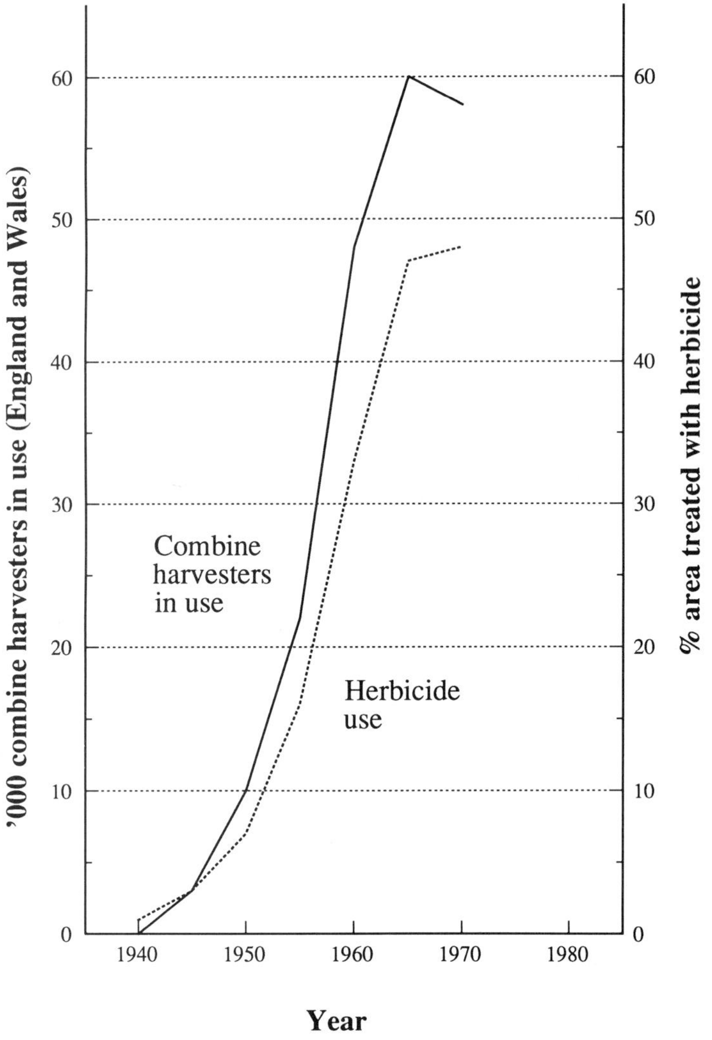

Fig. 16 — The growth of combine harvester use in UK cereal production in parallel with herbicide use.

During the eighteenth and nineteenth centuries, and the early part of the twentieth century, cereals were commonly grown in the UK and Western Europe as part of a rotation including one season's cultivation of a row root crop such as turnips.

Two seasons' consecutive cereal cropping tended to build up a mixed weed flora which was then 'cleaned up' by the frequent inter-row cultivations in the root crop. More than two seasons of cereals could not be tolerated because the level of weed infestation would become too serious a depressant to crop yield. The traditional Norfolk four-course rotation commonly used in nineteenth century England con-

Table 6 — First major herbicides for broad-leaved crops

Crop	Herbicide	Year of introduction
Soyabean	Trifluralin	1960
Cotton	Trifluralin	1960
Potatoes	Linuron	1960
Sugar beet	Pyrazon	1962

sisted of a root crop, barley, a legume crop and wheat, using a break crop between each crop of cereals.

The relative feed value per hectare of grain compared to fodder root crops exerted an economic pressure to increase the number of cereal crops in a cycle, if only the weed control problem could be overcome. Fodder roots in the early half of the 1900s produced about 1.6 t of starch equivalent per hectare while barley produced about 18t of starch equivalent. The resulting decline in fodder root production is illustrated by Fig. 17. The need for legumes to fix nitrogen was reduced by the greater use of fertilizer, and farming practice changed to give successive years of cereal production on the same land.

The range of weeds to become established in the cereal crops was very broad, including both monocotyledons and dicotyledons, but the relative importance of each species depended upon cultural practices, soil type and latterly, the selective effect of repeated herbicide use.

(ii) Effects of cultural practice on weed spectrum

The relative importance of different weed species is affected by the culture of the crop itself. Studies at the Stuttgart Agricultural College between 1870 and 1959 reported by Rodemacher showed that:

— Better soil nutrient conditions had eliminated certain species which were suited to poor, acid or alkaline soils.

— Use of cleaner seed samples at sowing time had reduced the importance of those species whose seed had been commonly distributed in impure crop seed, e.g. white mustard (*Sinapis alba*) and runch (*Raphanus raphanistrum*).

— Improved or more thorough cultivation of the stubble or seedbed had reduced the density of certain perennial species such as creeping thistle (*Cirsium arvense*), field bindweed (*Convolvulus marvensis*), and couch grass (*Agropyron repens*).

— The increased shading resulting from thicker, higher yielding, better fertilized crops had given a comparative competitive advantage to climbing species (such as cleavers (*Galium aparine*) and black bindweed (*Polygonum convolvulus*)), shade tolerant species (such as chickweed (*Stellaria media*)) or ephemerals† (such as *Poa annua*).

† Ephemeral species are those with a short generation interval and high fecundity which can rapidly colonize empty areas.

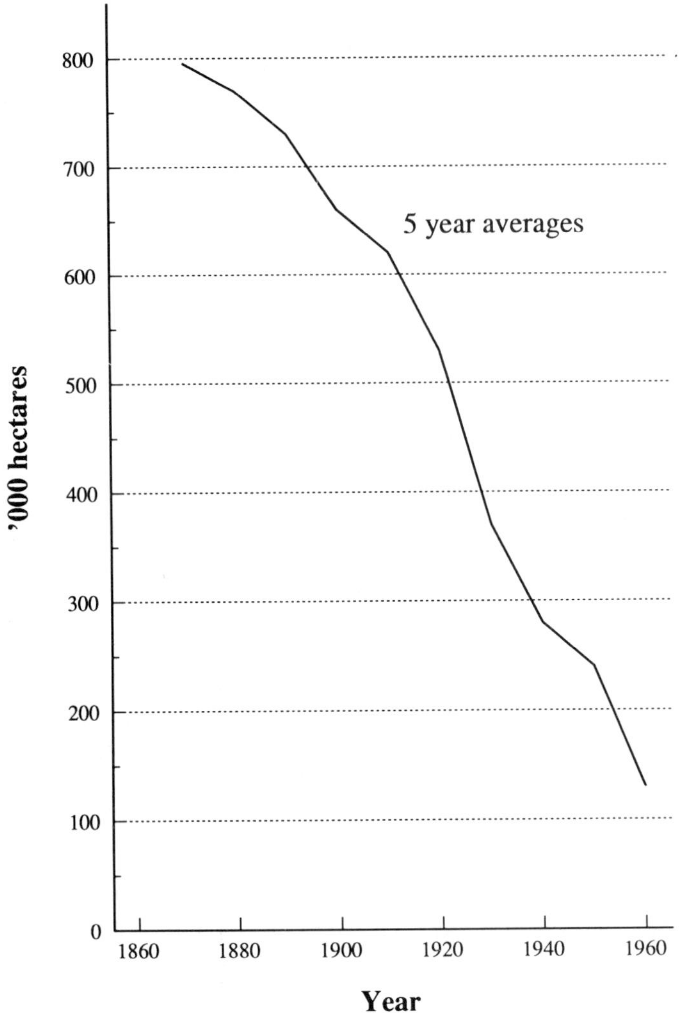

Fig. 17 — Decline in area of fodder roots grown in England and Wales. Source: MAFF.

— The increased use of mechanization and the need to plant spring sown crops as early as possible in the year brought their own influences to bear, removing an opportunity to remove first weed flushes by mechanical cultivation.

A significant mechanical influence was the advent of the combine harvester. When cereal crops were harvested in sheaves, either by hand or by binder, many of the weed seeds were carried off the field along with that season's weed plants in the sheaves. Combine harvesters return a proportion of weed seeds to the field together with the threshed straw and chaff and, because they cannot practically be cleaned thoroughly between fields (or between heavily infested and lightly infested parts of fields), they tend to aid weed seed distribution.

— The tendency towards monoculture, particularly continuous cereal growing, has provided an ecosystem favouring selection of those species most suited to co-existence with the crop.

(iii) The effect of herbicide use on the composition of weed flora

Individual herbicides tend not to be equally active against all weed species. There will be a tendency, therefore, for the most susceptible species to decline and the more resistant ones to predominate in a weed population.

Those species which show the greatest ability to survive herbicide programmes tend to have one or more of the following characteristics:

— Inherent physiological or morphological characteristics which reduce the chances of finding a chemical which will effectively select between the weed and the crop in which it occurs, for example wild oats (*Avena fatua*) in cereals.

— Very prolific seeding rates or very high levels of seed viability which allow the species to reappear in large numbers despite a severe reduction in the population in the previous year (e.g., *Avena fatua* in cereals and *Setaria* species in maize and soyabeans in North America).

— Large reserves of energy in large seeds, rhizomes or stolons which enable the plant to recover from initial chemical injury (e.g., many of the large-seeded weeds in soyabean such as *Ipomoea, Xanthium* and rhizomatous *Sorghum halepense*) or to grow from lower soil depths away from a surface layer of herbicide.

— Widespread or late germination times which require prolonged chemical activity or several applications to control all seedlings (e.g., *Avena fatua*).

— Long and widely varying seed dormancy periods which allow a species to appear in significant numbers irrespective of the control of that species achieved in previous years (e.g., *Avena fatua*).

— Species of high competitive ability which can germinate late and grow up through a heavily shaded crop canopy (e.g., *Galium aparine*).

(iv) The progressive coverage of weed spectrum by new compounds

Saunders in 1949 listed 38 weed species of importance in cereals in the UK, representing the flora prior to the effects of widespread herbicide usage. Of these, four were members of the family Gramineae (to which the cereals themselves belong). Two were annual grasses (wild oat (*Avena fatua*) and blackgrass (*Alopecurus myosuroides*)) and two perennials (couch (*Agropyron repens*) and water grass (*Agrostis stolonifera*)). These were to remain uncontrolled for some years to come. Of the other broad-leaved weed species, the majority were susceptible to the phenoxy-acid 'hormone' weed killers (2,4-D and MCPA) developed in the 1940s.

Of the minority of relatively resistant broad-leaved species, a few were particularly suited to the ecology of intensive cereals, notably stinking mayweed (*Anthemis cotula*), mayweed (*Matricaria* sp), cleavers (*Galium aparine*) and the Polygonums, knotgrass (*Polygonum aviculare*), redshank (*Polygonum persicaria*) and black bindweed (*Polygonum convolvulus*). These, in common with the grass weeds, constituted a target for new compounds to be used, either as panacea treatments or as additives for mixture with the hormones.

Plate 6 — Wild oats (*Avena fatua*) in wheat.

Plate 7 — Cleavers (*Galium aparine*) in wheat.

The inventions which appeared to solve the broad-leaved weed escape problem have, in fact, been used as additives in mixtures with phenoxy-acids. The major compounds which dominated the new mixtures were ioxynil and bromoxynil, introduced in 1963 by May & Baker, controlling *Anthemis cotula*, *Matricaria* sp and the *Polyganacae*, and dicamba (introduced by Velsicol in 1965) which has a similar spectrum, but includes *Galium aparine* (Table 7).

Table 7 — Effect of ioxynil/bromoxynil and dicamba on phenoxy-acid resistant broad-leaved weeds (Source: Fryer & Makepeace, 1978)

Major phenoxy-acid resistant broad-leaved weeds	Compounds	
	Ioxynil/Bromoxynil (1963)	Dicamba (1965)
Anthemis cotula	S	
Matricaria sp	S	S
Galium aparine	R	S
Polygonum aviculare	S	S
Polygonum persicaria	S	S
Polygonum convolvulus	S	S

R, resistant to control; S, susceptible to control by the chemical.

These compounds became particularly attractive commercially for inclusion in mixtures because of their very high levels of biological activity, allowing them to be incorporated in mixtures without inordinately increasing their cost. This issue will be discussed below but normal herbicide rates at this time averaged 1–3 kg of active ingredients per hectare while ioxynil and bromoxynil have been used at 0.2–0.8 kg/ha and dicamba at 0.1–0.2 kg/ha.

Grass weed problems were long accepted as a phenomenon of cereal growing, particularly in continuous cereals where phenoxy-acid herbicides were removing the broad leaved species and leaving an 'ecological niche'. Early experiments on Broadbalk at Rothamsted Experimental Station in the latter half of the nineteenth century showed that, even without 'artificial' selection pressures of herbicides, grass weeds become rapidly established. Continuous wheat growing started on the experimental plot in 1843. By 1852 wild oats (*Avena fatua*) had spread to the extent that teams of men and boys were employed to rogue out plants by hand. Black grass (*Alopecurus myosuroides*) also became a significant problem. These were the two annual grass species noted by Saunders in 1949 and both have provided a major research target for chemical control since the 1950s. Dadd estimated in 1956 that between 20% and 30% of fields in the Eastern counties of the UK had 'serious infestations' of wild oats.

The first solutions to wild oats appeared in 1958 with the introduction of barban by the Spencer Chemical Company as a treatment for use after the germination of both crop and weed and in 1961 when Monsanto introduced triallate as a residual herbicide to be applied before crop or weed emergence. Wild oats are a particularly difficult species to control because they have a wide range of germination times, have a long dormancy period in the soil (providing a seedbank for subsequent seasons) and produce large numbers of highly viable seeds. To prevent an increase in infestation between successive years, let alone achieve a reduction, control of the order of 98% is considered necessary. This is a very exacting requirement and has left scope for newer and better compounds to be discovered.

A number of compounds providing selective control of black grass appeared in the late 1960s. The first, nitrofen (from Rohm & Haas in 1964) was one of the least effective. Other notable compounds which have been introduced are terbutryne (Ciba-Geigy, 1966), metoxuron (Sandoz, 1968), methabenz-thiazuron (Bayer, 1968) and chlortoluron (Ciba-Geigy, 1969). This left the need for products which effectively combine good control of the range of both grass and broad-leaved weeds and of the perennial grasses (*Agropyron repens* and *Agrostis stolonifora*).

In 1975 Hoechst launched isoproturon, aimed at the market for total control of all annual weed species. There has still been scope for better annual grass weed herbicides for wheat and barley to give broader spectrum of weed control and more flexibility in timing of application. Recent introductions have included fenoxaprop-ethyl (Hoechst), tralkoxydin (ICI) and imazamethabenz (American Cyanamid) in the late 1980s.

The control of perennial grasses has been tackled outside the growing season either by post-harvest stubble cleaning or by sprays before planting in the spring. Many of the treatments have involved a combination of mechanical cultivation and chemicals. The problem has been to find a chemical potent enough to kill the perennial grasses without leaving phytotoxic residues to damage the crop when subsequently planted. Paraquat (ICI) and glyphosate (Monsanto) have both succeeded in this as they are deactivated by the soil and thus leave no harmful residues.

(v) Product evolution in US soyabeans

An analysis of the evolution of herbicides for soyabeans in the USA produces a similar picture to that just described for UK cereals. Before the 1950s there were no chemical control methods for weeds in soyabeans. Normal practice was inter-row cultivation. In the early 1950s dinoseb was marketed by Dow as a contact herbicide for use prior to crop emergence. Timing of application was critical to within a few days and the level of weed control was only moderate. Neither this compound, nor the others tried on soyabean during the mid-1950s were sufficiently effective. Allidochlor, a very short persistence residual, only controlled grass weeds and its lack of crop safety limited it to high organic matter soils. Naptalam was very dependent upon levels of rainfall and fineness of soil texture and also posed serious crop phytotoxicity risks. Chlorpropham had too limited a spectrum of weed control.

The early 1960s marked the breakthrough in chemical discovery. Chloramben (introduced by Rorer Amchem in 1960) was the first compound to give good control of a range of major grass and broad-leaved weeds. It was safe to the crop and consistent in its performance in the midwest of the USA (the major soyabean growing area) irrespective of soil tilth and weather.

Its outstanding efficacy resulted in very rapid growth of sales (Fig. 18). It had, however, a deficiency in its control of large seeded broad-leaved species, notably *Xanthium* sp, *Abutilon, Polygonum, Datura* and *Ipomoea*. It also performed poorly on light sandy soils.

In 1962 attempts were made to use 2,4-DB (a member of the phenoxy-acid group of herbicides) to control *Xanthium* post-emergence but this failed commercially because of the very critical application timing required for safety to determinate varieties of soyabeans (10 days before flowering until mid-bloom) and general phytotoxicity to indeterminate varieties.

In 1963 trifluralin was introduced by Eli Lilly as a very safe, reliable and broad-spectrum treatment. It was the only serious competitor to chloramben and was introduced at a lower price per hectare ($8.6 compared to $9.6). It had similar weaknesses in control of large seeded broad-leaved weeds. Its efficacy and early price competitiveness resulted in similar rapid growth of sales to that experienced by chloramben (Fig. 18). A further advantage was its role as the first truly effective herbicide for cotton (for which it is now the standard treatment), the alternative crop to soyabean in the rotation programme in the southern states.

During the period from 1963 to 1969 a number of similar compounds, showing only minor advantages or disadvantages, were launched, but failed to make significant inroads. Vernolate (from Stauffer) was very similar to trifluralin, giving better control of *Cyperus* sp but posing a greater risk of crop damage.

Nitralin (from Shell) was similar to trifluralin on sandy or low organic matter soils but poorer on the higher organic matter soils of the corn belt. Attempts to establish commercial or technical advantages over trifluralin failed and the compound has been withdrawn. Linuron (from Du Pont) gave better control of *Echinochloa* and *Eleusine* than chloramben and of *Polygonum* and *Abutilon* than trifluralin but was more phytotoxic to the crop than either. Rates are critically dependent on soil texture and organic matter content, the higher organic matter soils being more suitable. Soil moisture and post-application rainfall are also critical.

In 1969, the next major compound appeared, alachlor from Monsanto. Alachlor offers a good spectrum of control of grasses and small-seeded broad-leaved weeds and is very safe to the crop. It offered the first pre-emergence, non-incorporated treatment. (The previous compounds have to be incorporated into the soil surface with a light cultivator before planting the soyabeans.) Its success has probably been enhanced considerably by its importance as a herbicide for maize, the major alternative crop to soyabeans in the rotation in the major corn-producing states, and its ability to produce an effective treatment in mixture with linuron.

The herbicide progamme, however, entered the 1970s without a satisfactory control of large-seeded broad-leaved weeds or perennial grasses. This is the analogous situation to that described for cereals, which entered the 1960s without a satisfactory control of grass weeds (notably wild oats and black grass). There was also a need to control grass weeds which escaped from the pre-emergence treatments.

These became major targets for research by major pesticide companies. Two significant compounds for large-seeded broad-leaved weeds were introduced in the early 1970s. Metribuzine, introduced by Bayer in 1973, offered very good control of the more difficult, large-seeded broad-leaved weed species but still showed weakness on *Ipomoea*. On heavier soils it is very active (0.25–0.75 kg ai/ha compared to 2–4 kg/

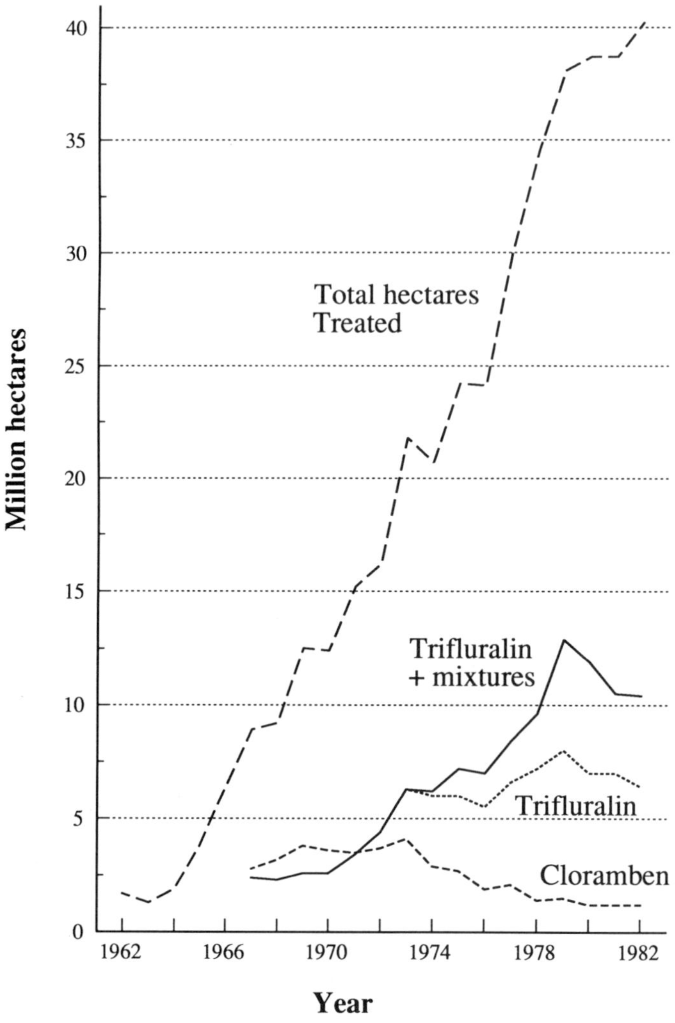

Fig. 18 — Early growth of use of effective herbicides in soyabeans in the USA. Source: Doane.

ha for alachlor). On lighter soils there are crop phytotoxicity risks which can be minimized by mixing at very low rates with trifluralin or alachlor, but at the expense of some loss in efficacy on difficult species. Bentazon, from BASF in 1974, offered the first safe chemical for use as an overall spray over an established soyabean crop to control escapes of the large-seeded broad-leaved species. It is excellent on the major species *Xanthium* but leaves scope for improvement elsewhere.

During the 1980s a further series of products was introduced to provide better control of the large-seeded broad-leaved weeds. These included the diphenyl ether compounds acifluorfen (Rhône-Poulenc and Rohm and Haas) and fomesafen (ICI) which give good control of difficult weeds such as *Abutilon* sp, *Xanthium* sp and *Ipomoea* sp at rates of 0.25–0.5 kg ai/ha at early growth stages of the weeds.

The 1980s also saw the introduction of herbicides for post-emergence control of perennial grasses and grass weeds which escaped the pre-emergence treatments with, for example, trifluralin or alachlor. These products included fluazifop-butyl (ISK/ICI), and sethoxydin (Nippon Soda/BASF), quizalofop-ethyl (Nissan/Du Pont), and alloxydin (Nippon Soda/BASF).

Plate 8 — Weed control in soyabeans showing an area treated with herbicies (left) compared with an area infested with giant foxtail (*Setavia faberii*).

(vi) Evolution of the US maize herbicide market

The third example of evolution of herbicides is taken from the world's largest herbicide market, maize (or corn) in the USA.

Before the development of 2,4-D in the mid-1940s, weed control in maize relied upon mechanical cultivation. 2,4-D, however, suffered from a number of disadvantages which opened the way for a new generation of broader spectrum residual herbicides. It provided no control of grass weeds, had no residual life in the soil to provide a prolonged period of weed control and could cause damage to the prop roots of the maize crop if applied when the crop was 8–10 in (20–25 cm) high. Unfortunately, this coincides with the most effective time for application to control a

(a)

(b)

Plate 9 — Weed control in maize. (a) Young crop competing with weeds. (b) Weed-free crop treated with herbicides.

major dicotyledonous weed, *Xanthium* (cocklebur). Much of the use of 2,4-D now is as an aesthetic clean-up, in sequence with grass killers.

The following chronological account, to be considered with the product growth graph, Fig. 19, illustrates the series of technical advances in weed control achieved since 2,4-D and their impact on the US market.

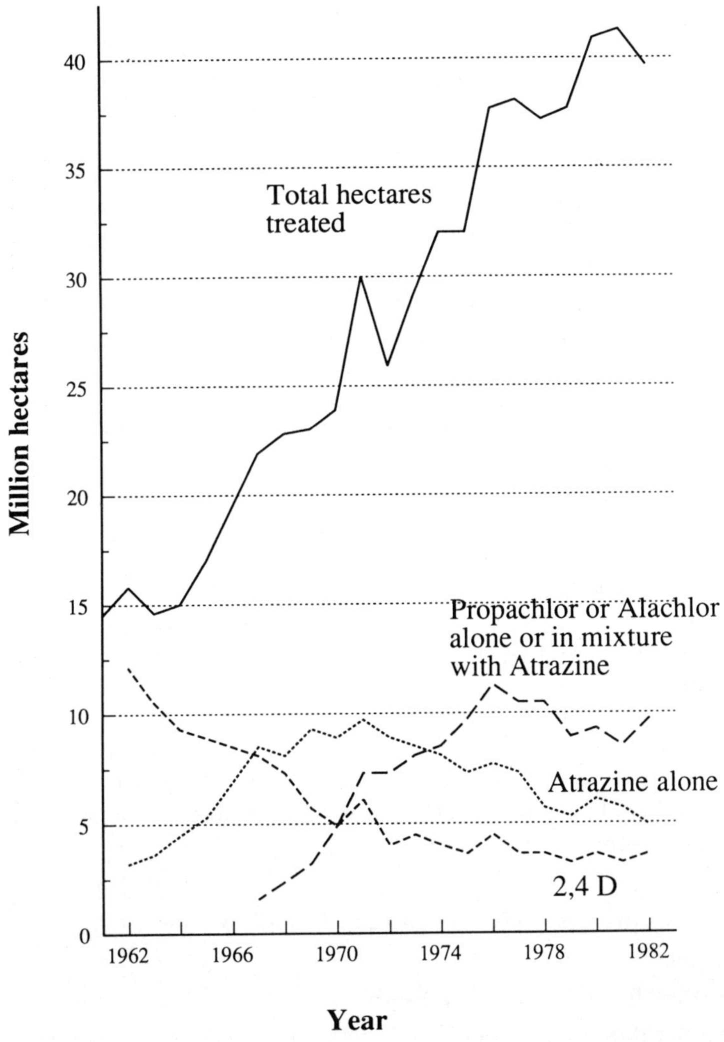

Fig. 19 — Early growth of effective herbicides in maize in the USA.

Atrazine was introduced by Ciba-Geigy in 1958 as the first broad-spectrum residual herbicide for maize. It is very safe to the crop (it possesses an inherent physiological selectivity due to the crop's ability to detoxify the compound) and reliably controls a wide range of both grass and broad-leaved weeds. It is, however, weak against a number of common grass weeds, notably *Digitaria* sp, *Panicum dichotomiflorum, Sorghum halepense* and *Sorghum bicolor*, and also on *Cyperus* sp and *Asclepias* sp (a broad-leaved perennial weed). It also has efficacy problems on soils with high organic matter contents (over 5%) where high application rates are required; these can leave residues in the soil which affect subsequent crops. Its wide general appeal and effectiveness, however, have made it the basis for weed control programmes in maize in the USA.

The weaknesses on a number of common grasses set the stage for the introductions of the middle to late 1960s.

Propachlor, introduced by Monsanto in 1965, is more effective on those grasses which show a measure of tolerance to atrazine and also offers a fair breadth of spectrum for use on the heavy organic soil.

Alachlor, a very close chemical analogue of propachlor, was introduced, again by Monsanto in 1966, as a potentially better (and more profitable) version of propachlor, being effective at about half the application rate per hectare and having a longer period of persistence (10–12 weeks compared to 4–6 weeks for propachlor). Propachlor remains on the market, probably because of its advantages on very heavy soils, for example in North Central Iowa, Southern Minnesota and East Central Illinois.

Butylate was an attempt by Stauffer to enter the new market segment for grass killers to supplement atrazine. It is a chemical from a group of thiocarbamates on which the company had been working for some time. It was not as generally effective as alachlor and has made considerably less impact. Its main distinguishing feature is better control of *Cyperus* sp.

Cyanazine, launched by Shell in 1971, was aimed to fill the market segment for general residual weed control (the atrazine effect) under those conditions where the atrazine residue problems ('carry-over effects') exist. It has a similar spectrum but shorter residual life.

Given the existence of good grass control compounds and a very good general compound, weak on certain grasses, the logical development has been for mixtures of atrazine and grass killers to give excellent overall results. The use of atrazine mixtures has also enabled a reduction in atrazine application rates, with a commensurate reduction in carryover problems.

(b) Product improvement in insecticides and fungicides

(i) Insecticides

Major improvements in insecticide performance have resulted from:-

— Increased duration of control offered by the introduction of systemic products. For example, the systemic granular insecticide carbofuran can be used on paddy rice to control a spectrum of rice pests over a period of time equivalent to that achieved by two or three sequential applications of non-systemic products.
— Control of pests which have become resistant to the previously available compounds. The phenomenon of pest resistance is the result of a progressive evolutionary pressure in favour of the more resistant members of a pest species in the presence of pesticidal compounds. Because there is a wide range of suscepti-

bilities to a given chemical treatment within a species, resulting from inherent genetic variability, there are usually treatment survivors for whom a normal dose rate of chemical is sub-lethal. Those survivors then tend to pass on chemical tolerance characteristics to subsequent generations, which, in turn, require a higher chemical application rate to achieve a given percentage kill. Progressively, dose rates have to be increased and, in certain situations, this reaches a level at which control becomes totally ineffective.

An illustration of the progressive build-up of resistance was given by Reynolds (1976) quoting the multiplication in dosage rate of methyl parathion required to control *Heliothis virescens* (the tobacco budworm) on cotton in Texas and Mexico (Table 8).

Table 8 — Comparative resistance of the tobacco budworm (*Heliothis virescens*) to methyl parathion[a]

Location	Year	Increase in resistance
College Station, Texas[b]	1964	Susceptible
College Station, Texas[b]	1968	4×
College Station, Texas[b]	1969	6×
College Station, Texas[b]	1970	14×
College Station, Texas[b]	1971	11×
College Station, Texas[b]	1972	51×
Weslaco, Texas[b]	1968	5×
Weslaco, Texas[b]	1969	11×
Weslaco, Texas[b]	1972	25×
Tampico, Mexico[c]	1969	46×
Tampico, Mexico[c]	1970	61×
Tampico, Mexico[c]	1970	201×

[a] Adkisson, compiled by Reynolds (1976).
[b] Data from Nemac & Adkisson, compiled by Reynolds (1976).
[c] Data from Wolfenbarger *et al.*, compiled by Reynolds (1976).

Cotton producers in the USA have, in fact, been beset by problems of insect resistance over a range of major species and pesticidal chemical groups. The problems began in the mid-1950s when the boll weevil (*Anthonomus grandis*) first developed resistance to organochlorine insecticides, and, shortly afterwards, *Heliothis virescens* started to show organochlorine resistance as well. To combat these resistance problems, a very effective new group of compounds, the organophosphates (OPs) (principally methyl parathion but also azinphos-methyl, EPN, malathion and others) was introduced. These compounds had to be applied relatively frequently because of their short residual activity and they were also very harmful to predatory species in the insect complex, predators which tended to keep the pest species in check.

As noted above, *Heliothis* resistance developed very rapidly and, in the absence of natural predators, produced serious and economically damaging outbreaks. Farmers were in a dilemma. Without OP insecticides, damage caused by boll weevils was too great to enable economic yields to be produced. With OPs, the destruction of *Heliothis* predators allowed *Heliothis* populations to rise to the point at which production was again jeopardized.

The problem was ameliorated again by a programme of late autumn chemical and cultural practices to control boll weevils just prior to winter diapause, allowing the following season's spray programme to be delayed and allow some build-up of predator populations. A new chemical group, the synthetic pyrethroids, has since replaced the organophosphates. Care is being taken to monitor pyrethroid resistance risks and to manage spray programmes in the light of resistance knowledge.

The development of resistance by two major cotton pests described briefly above is purely illustrative of a general problem of the increasing resistance of a wide range of species to the limited range of chemical groups so far discovered. Table 9,

Table 9 — Number of species of insects and mites in which resistance to one or more chemicals has been documented

Year	Species
1908	1
1928	5
1938	7
1948	14
1954	25
1957	76
1960	137
1963	157
1965	185
1967	224
1975	305
	(+59 unconfirmed)

Data from various sources, compiled by Georghiou and Taylor (1976).

compiled by Georghiou and Taylor (1976) shows the increase in numbers of species of insects and mites in which resistance to one or more chemical has been documented.

Considerable advances have been made in the study of the biochemical and physiological mechanisms of resistance, especially in arthropods, and in understanding the biochemical characteristics of specific defence mechanisms and insect genetics.

As one would expect, the rate of development of resistance is a function of the biochemical and physiological modes of action of pesticidal compounds (their entry into the insect, transport to active site, binding to the active site and the natural biochemical process which they then disrupt), the frequency of occurrence and nature of the genes for resistance in the population, the generation interval (the more generations per year, the faster resistance occurs), the selection pressure imposed by pesticide use, and the extent to which those pressurized are diluted by migratory rather than inbreeding populations.

The situation is further complicated by cross-resistance between chemicals of different chemical groups but with related (in some way) modes of action.

(ii) Fungicides

Major improvements in fungicidal performance have been achieved; improvements in quality, spectrum and duration of control. Before the late 1960s, most of the available fungicide compounds, from the simple inorganic compounds such as sulphur and copper to the range of major thiocarbamate compounds, were non-systemic protectants. The compounds would be deposited on the leaf surface and would kill superficial fungal mycelium, prevent the production of new fungal spores (thereby reducing the spread of disease within the crop canopy) or prevent spore germination. They would not enter the plant tissue or move within the plant.

The very mode of action of these compounds limits their efficacy for a number of reasons. They cannot control disease developing within the plant tissue, limiting their effectiveness on non-superficial diseases. This means that treatment must be preventative and made *before* the disease has become established. Plant growth also causes protective chemical coatings on expanding surfaces of leaves,stems or fruit to become ruptured, leaving unprotected sites for infection. Also new tissue developed at growing points is unprotected.

An advance since the late 1960s has been the advent of xylem systemic compounds — chemicals which move in the plant xylem transport vessels, reach the fungal mycelium within the plant tissue and protect new areas of growth. This gives control which is better and of longer duration. Xylem systemic products fall short of a total systemic activity and there remains scope for finding totally systemic products which also exhibit phloem mobility.

It has been postulated within the industry that research into applications of better systemic fungicides in turn throws up new opportunities for their use not previously recognized. Some historical evidence for this aspect of 'science push' creating an awareness of a demand which had previously been latent, is the development of markets for fungicides on cereals (wheat and barley) in Western Europe.

It was recognized for some time that wheat and barley in Westwen Europe was attacked by fungal diseases, for example, cereal mildew (*Erysiphe graminis*), rusts (*Puccinia striiformis* and *Puccinia hordei*) and eyespot (*Psuedocercosporella herpotrichoides*). Commonly grown varieties were, however, bred for disease resistance and the yield loss from the residual level of infection was thought to be small until candidate systemic fungicides were discovered and tested. The earliest of these were ethirimol from ICI and tridemorph from BASF. Field tests showed that fungicidal application could increase yields by about 10% under average conditions.

Plate 10 — Wheat powdery mildew (*Erysiphe graninis*).

Gradually the use of ethirimol and tridemorph created an awareness of the technical benefits of cereal fungicide use. The economic climate for cereal growers under the protection of high guaranteed prices within the European Economic Community gave incentives to move to high input/high output production systems, using high levels of fertilizer and pest control chemicals to maximize yield. The stage was set for a spate of new fungicide products to give better disease control and to encompass the wider range of diseases that then became recognized. By the mid-1980s the range of foliage and ear diseases that could be controlled by a season's spray programme had expanded from mildew and rust to the following list:

Wheat:	Powdery mildew	(*Erysiphe graminis*)
	Brown rust	(*Puccinia recondita*)
	Yellow rust	(*Puccinia striiformis*)
	Eyespot	(*Pseudocercosporella herpotrichoides*)
	Sharp eyespot	(*Rhizoctonia solani*)
	Septoria	(*Septoria tritici*)
	Septoria	(*Septoria nodorum*)
	Fusarium	(*Fusarium* sp)
Barley:	Powdery mildew	(*Erysiphe graminis*)
	Brown rust	(*Puccinia recondita*)
	Net blotch	(*Pyrenophora teres*)
	Rhynchosporium	(*Rhynchosporium secalis*)

The number of products and mixtures of products has also grown to meet the new market demand. These include, for example, the series of triazole-based compounds listed in Table 10, which possess broad spectrum systemic activity. These have been

Table 10

Compound	First announced	Company
Triadimefon	1973	Bayer
Diclobutrazol	1977	ICI
Prochloraz	1977	Schering/FBC
Propiconazole	1979	Ciba-Geigy
Flutriafol	1983	ICI
Flusilazol	1984	Du Pont

used in mixture with protectant compounds such as captafol, chlorothalonil and mancozeb to improve the range of diseases controlled and with the systemic carbendazim to control eyespot (*Pseudocercosporella herpotrichoides*) as well as other later season diseases.

Although application of fungicides to the American soyabean crop appeared to present an analogous technical opportunity, this has failed to develop because of low crop values. Du Pont, makers of one of the earliest broad-spectrum systemic fungicides, benomyl, were seeking new uses for their product. Soyabeans had been observed to develop some disease symptoms at the very end of the season but this was previously considered to be unimportant and part of the syndrome of crop senescence. University plant pathologists working with benomyl in the Southern USA discovered that late season applications of fungicide not only delayed the senescence symptoms by controlling a fungal complex, but also achieved about 10% greater yield of beans by allowing an extra period of photosynthetic activity late in the season. Growth of this use, however, has been limited compared to that use of cereal fungicides in Europe, because of the poor economics of the crop.

The advent of systemic compounds has, however, created new problems. The older protectant compounds attack a range of fungal metabolic processes and do not appear to leave sufficient survivors of sub-lethal treatment to encourage resistance. It appears, from experience with systemics so far, that these compounds tend to be more specific in their mode of action (possibly an essential feature to imbue them with selectivity between fungal and plant phytotoxicity within plant tissue) and they are present within plant tissue for certain periods at sub-lethal doses following both their dilution by plant growth and partial degradation. These sub-lethal doses create the potential environment for diseases to develop resistance — a phenomenon which is now becoming apparent with several classes of systemic compounds.

(c) Reduction of treatment cost

The development of the pesticide business until the 1980s has tended to be orientated towards improvement in product performance because of the perceived scope and potential demand. However, as the level of technical efficacy and availability of products has risen, and the number of potential alternative treatments for each pest control need has proliferated, increasing competitive pressures have led to cost-reducing innovations. These pressures have been brought to bear by farmers looking for the most cost-effective treatment. Product distributors have sought those products with the lowest cost to enable them to sell at competitive prices but retain worthwhile distributive margins. Chemical manufacturers have attempted to gain or maintain market share in the face of strong competition.

Cost reductions arise in a number of different ways:

(1) For existing compounds through economies of scale and progressive improvements in production process technology.
(2) Development of new compounds which require much lower rates of application per crop hectare, or require fewer applications per crop hectare per season.
(3) Development of low-volume application systems which, although they apply the same quantity of active chemical per hectare, require lower volumes of total spray solution and are, thereby, easier and cheaper to apply — a particular advantage when water cartage is expensive, e.g. for aerial application.

The history of the industry has examples of order of magnitude increases in biological efficacy (Table 11) but it has not been possible to assemble these into any very clear steady progression, of the type investigated by Bright (1968), from which to predict the rate of progress to be expected.

Examples of compounds requiring fewer applications per season include carbofuran (an FMC insecticide) on rice and the latest generation of triazole fungicides which, in two or three applications per season, can control powdery mildew as well as 8–10 applications of sulphur.

(d) Product serviceability — the environmental factor

The widespread expansion in the use of agricultural chemicals has been accompanied by an increased awareness of the need to guard against deleterious side-effects.

Table 11 — Increases in pesticidal potency of products

	Date introduced	Rate
Herbicides		
Sodium chlorate		100–200 kg ai/ha
Most currently used selectives	1950–present	1–3 kg ai/ha
Diphenyl ether	*c* 1980	0.2–0.5 kg ai/ha
Sulphonyl ureas	*c* 1985	0.01–0.05 kg ai/ha
Fungicides		
(Exemplified by rates required for apple mildew control)		
Sulphur	1848	1500–6000 ppm
Benomyl	1967	250 ppm
Fenarimol, Etaconazole, Hexaconazole	*c* 1980	10–20 ppm
Insecticides		
Most commonly used insecticides	1940–1960	1–1.5 kg ai/ha
First generation synthetic pyrethoids (e.g. permethrin)	1973	0.05–0.1 kg ai/ha
Fully resolved synthetic pyrethroids (e.g. Deltamethrin, and Cyhalothrin)	1974	0.005–0.02 kg ai/ha

Users of chemicals should not be exposed to unacceptable hazards during spray applications. Residues in food should not put the consuming public at risk. Economic benefits derived from the use of chemicals as aids to productivity must be set against environmental risks resulting from the presence of new biologically active molecules in the environment.

In order to establish the extent of possible side-effects from agricultural chemical use, and to ensure that degrees of risk are not 'unacceptable', national governments have evolved increasingly stringent toxicological and environmental testing protocols which chemical companies must satisfy before gaining the legal registration approval needed before a product can be marketed for commercial use. In many countries a certain proportion of the toxicological and residue data is required before large scale field testing involving the sale of treated crops is permitted. One effect of the stringency of the legal registration requirements has been to increase the cost of new product R & D. This is discussed in more detail in Chapter 5. Another effect has been the banning of products currently marketed (e.g., the banning of most organochlorine insecticides, including DDT, in the USA on environmental grounds) and the institution of re-registration procedures for compounds which have been registered on the basis of earlier and less stringent protocol requirements, forcing manufacturers of existing products to bring the package of toxicological and environmental data on those chemicals up to the latest standards and to demonstrate that the economic benefits from use of the products significantly outweigh the risks.

Inevitably the effect of these proceedings will be the further banning of compounds which are either shown to be harmful or do not generate sufficient profit to the manufacturing company to warrant their being defended. Opportunities are thereby created for new products of demonstrated acceptability to be developed and marketed in sectors which may previously have been technically satisfied.

4

The agrochemical R & D process

OVERVIEW

In Chapter 3 we saw how the chemistry of agricultural chemicals evolved from its early beginnings in the chance discoveries of individual scientists and inventors. As the agrochemical industry has grown in size and complexity (a topic discussed in more detail in Chapter 9) the R & D process leading to new agrochemical products has become more complex and more highly structured.

The process is one which seeks to move as effectively as possible from the initial search for new highly effective molecules in novel areas of chemistry through to a position where a single molecule has been selected and sufficiently well characterized for it to be practically applied as an agrochemical. Before a molecule can become a commercial agrochemical, very detailed information must be discovered about:

— its biological properties, to enable reliable recommendations to be made for its use;
— its safety to users, food consumers and the environment;
— optimum methods of manufacture and formulation, to achieve an acceptable, cost-effective product.

The search, selection and characterization stages for each new product tend to follow a common pattern, although the detailed problems will differ greatly from chemical to chemical. The stages can be broadly classified as:

— targeting,
— synthesis,
— screening,
— evaluation,
— development,
— post-sales technical support.

The targeting stage determines what biological effects will be sought and which chemicals will be made and tested. A very large number of chemicals are then tested for biological utility in the hope that a proportion will show properties which are sufficiently attractive to justify further testing. The subsequent cascade of tests will generate progressively more information about each molecule tested and will whittle down the numbers of molecules meriting progress to the next stage of testing. We will consider the detailed nature of this process step by step.

TARGETING

Science has not yet advanced to the point at which a new agrochemical can be designed to meet a predetermined set of specifications. The majority of chemicals tested as potential agrochemicals have no practical utility. On average, only about 1 in 10 000 of the original candidates ever becomes a commercial product. However, it is vital that the effort devoted to synthesis of novel chemicals and to the subsequent series of tests is given the maximum opportunity to produce economically worthwhile results. Chemical companies investing in agrochemical R & D are likely to need to spend $30m to identify a molecule with the right characteristics to be worth developing as a new product and a further $30–60m to characterize that product sufficiently for it to be sold and used commercially.

Research targeting aims to enhance the efficiency of the process by clarifying objectives and helping to ensure that the emphasis of synthetic programmes and testing sequences reflects these objectives.

Targeting will have chemical, biological, agronomic, environmental and economic dimensions and varying degrees of generality or precision.

Each company has to decide what total level of resource to commit to agrochemical R & D in relation to its perceptions of the long term financial opportunity likely to be offered by new products and in relation to shorter term constraints imposed by its physical organization and cash flow projections (i.e. what it can afford). Within that overall commitment it must apportion resources between areas of activity, such as chemistry, biological testing, environmental work, toxicology, process development, formulation, and disciplines, including herbicides, fungicides, insecticides, plant growth regulators, and exploratory biosciences. Within each discipline, it then has to set priorities on the type of product it is trying to produce for markets which will begin after about 10 years and peak in 20 years or later. As the market matures and the demands expand, there is an increasing need to defend and re-register existing commercial products, leading to increased competition between work on existing products and the development of novel products.

In forming its list of targets for research, research management has to take into account:

- projected trends, within the existing business, of product efficacy (e.g. rates of use, required spectrum of pests controlled and duration of control), scale of sales and pricing trends;
- the existence of unsolved problems, or the likely emergence of new unsolved problems (e.g. resistance of pests, weeds or diseases to existing chemicals);
- trends in world agriculture, both technical and economic;
- trends in legislative constraints, driven by concerns for user and environmental safety and by the effects of political lobbies;

— trends in public sector and academic science;
— competitive research activity;
— commercial competition in the market.

It is not the purpose of this book to display alternative futures, although some crystal ball gazing takes place in Chapter 10. Each company will have its own and often different views. Very marked differences in perception were evident from the different strategies apparently being pursued by different major agrochemical companies in the mid-1980s. Some had very specialist approaches e.g. concentrating on herbicides, while others covered a broad area of crop protection chemistry. Some put a large proportion of effort into novel areas such as plant growth regulators while others concentrated on conventional pesticides.

Within the broad target areas set, there will be finer specifications.

Chemical targeting will define the areas of chemical synthesis upon which the corporate chemists will concentrate. These will reflect a balance between random, 'hunch following' and 'lead following' synthesis; a subject considered in more detail in the next section.

Safety targeting will set parameters for, for example, toxicity, environmental persistence or mobility in soil. Biological targeting in the 'conventional' pest control disciplines of herbicides, insecticides and fungicides will set priorities on certain crops and pest species and will set performance targets for level of activity, duration of control, mode of action, crop safety and other parameters.

Plant growth regulator targets may be set in terms of the biological response sought and the economic utility of that effect. 'Biological response' targets would be defined in terms of, for example, growth retardation, a flowering response or abscission or in terms of a range of biochemical, physiological or morphological effects deemed likely to lead to improvements in crop yield or quality or to aid crop response to stress.

'Economic utility' targets would reflect attempts to improve agricultural output or reduce production cost and would be defined in terms of perceived user needs such as cereal lodging control, tree growth containment, harvest aids or yield improvements in particular crops. Some historical examples were given in Chapter 1.

The degree of generalization or precision of the targeting will be reflected in the subsequent steps in the invention process between synthesis and its accompanying biological screen to indicate useful activity, and later stage 'practical' trials.

For any new agrochemical, R & D target priorities will be influenced by:

— perceived probability of success,
— ability to construct a suitable screening process,
— perceived value of the target product, should it be successfully invented and developed.

For conventional pesticides, some very rough estimates of chance of success will be available from precedent and experience. Testing sequences can be closely related to the target and targets can be valued with some degree of accuracy (at least within an order of magnitude). For novel areas of research such as plant growth regulators or agricultural application of 'biotechnology', targeting requires much greater insight to assess chances of success, to invent new testing procedures or to perceive and quantify a latent novel market demand.

SYNTHESIS

Molecules tested as potential new agrochemicals cannot yet be designed on a drawing board, or a computer screen, to meet desired specifications of performance and safety. The process still relies on serendipity, albeit with increasing help from science to improve the chances of the synthetic chemist looking in areas of chemistry where valuable molecules could be lurking. Many of the early products were found by chance through the random testing of chemicals which had been made for other purposes. Others were found by following a train of thought, as exemplified in Chapter 3 by the discovery of the organomercurial seed treatments and the 'hormone' weedkillers.

The modern supply of chemical compounds for screening as potential agrochemicals will contain many which have been made for other reasons (e.g. pharmaceuticals or industrial chemicals) but a large and increasing proportion will be new compounds, synthesized specifically as possible agrochemicals.

The laboratory synthesis of new potential agricultural chemicals is driven by a number of sources of ideas. These will include:

— Molecular structures chosen to mimic known active chemicals, either naturally occurring or from previous synthesis;
— Molecules invented following ideas derived from an understanding of biochemical processes and target enzyme systems;
— Chemical leads which arise from observations made during the testing processes. These may at one extreme be leads derived by the random testing of chemical compounds made for other purposes and, at the other extreme, a molecular structure derived from a process of structural optimization within a chemical group as a result of detailed structure/activity analysis and computer modelling.

Synthetic leads for plant growth regulators are likely to be derived in a similar way to those for conventional pesticides.

Considerable success has been achieved by chemical imitation around molecular structures previously shown to have activity. The companies who first discovered the active chemical area have often 'struck a rich seam' and generated a family of products. In areas where the range of active variants on a theme has not been comprehensively patented, there have also been rich pickings for competitors who follow suit. Often the number of active variants is sufficiently large and the range of structural variants so diverse that it is not possible for any one entrant to the chemical area to synthesize and characterize a sufficient cross-section to capture the whole patent territory. This is illustrated by examples for triazole fungicides in Appendix 1.

It is interesting to note that not only have many companies been successful in the same general areas, but chemical areas have often yielded new useful molecules 10 or 20 years after the area was first discovered. Many useful molecules have also arisen from bringing together, deliberately or fortuitously, parts of successful molecules from different series.

However, imitative chemistry does not necessarily give rise to imitative biology. The biological properties of close structural analogues can be very different. Changing one chemical group on a structure can totally remove activity or can confer undesirable properties.

Once a company has established biologically useful leads within a chemical class, biochemistry, physical chemistry and computer aided structure–activity analysis has markedly speeded the process of trial and error to arrive at the optimum structure.

Detailed analysis of the effects of changes of parts of a molecule on its physical properties helps to predict the optimum configuration to give the required uptake and movement properties for the preferred compound. Once a number of analogues have been made within a chemical group and tested for biological performance, it is often possible to build a predictive mathematical model which relates physicochemical measurements to biological effects. This enables more efficiently targeted synthesis and testing within the wide range of potential molecular configurations which exist within the broad structural class.

Parameters which appear important in affecting activity are hydrophilicity, electron charge distribution and stereo configuration (3-D shape). Other parameters which have been explored to less advantage so far include hydrogen bonding and measures of distance between critical atoms or centres.

Hydrophilicity/lipophilicity, a measure of the relative solubility in water and fats, affects the movement of compounds into and through different tissues and their transport to the site of action. Measures used are log P (the octanol : water partition coefficient of the whole molecule) and π (the partition coefficient of a substituent (e.g. Cl, CH_3) compared with hydrogen ($\pi_H=0$)).

If a mathematical relationship has been established between log P and biological performance with a set of examples from a group, it is simple to calculate the log P of as yet unmade compounds and to concentrate synthesis on those with the greatest chance of success.

The effect of adding different substituents on charge distribution can also be calculated.

Understanding the biochemical mode of action and the physical relationships between active molecules and an identified binding site on a target enzyme can help to optimize intrinsic activity. Better understanding of the structure of active binding sites and of the 3-D shape of molecules is now possible with the use of 3-D computer graphics as an aid to molecular design.

The advances in science required for such analyses to become possible have improved the chemists' decisions on which molecules to make when following a very close lead, but are not yet able to specify the definitive optimum.

PATENT PROTECTION

Each new agrochemical represents very large investment of skill and financial resource and the taking of large financial risks. For that investment to be adequately rewarded, the inventors and developers require some protection from competitive activities of other organizations, which could take the completed invention and exploit it without having to bear the development costs and risks themselves. That protection is provided in part by patent rights taken out by the inventors of the chemical itself (if it is a novel compound not previously made and cited in literature) or on its 'unexpected' agrochemical utility. Patents are normally applied for to protect specifically defined classes of chemical within which a number of examples have been made and tested.

An example of a chemical area which could be covered by a specific patent could be compounds of the structure

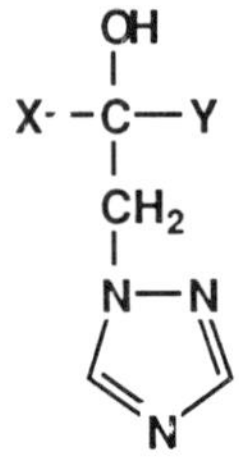

Where Y can be phenyl or an alkyl chain of 2–8 carbon atoms, X can also be phenyl or an alkyl chain of 2–8 carbon atoms. The phenyl ring may be unsubstituted or substituted with the halogen chlorine or fluorine in one or more positions. The alkyl chain may be straight or branched. Examples of analogues are given in Fig. 20.

Patent applications will normally be filed as an initial 'priority filing' as soon as a novel chemical group is shown to have activity in the first screening tests. Inventors are then given a period of one year to make and test sufficient examples to 'prove' their claims.

At the end of this period, the inventors will file their completed set of data in support of the claim, excluding areas which cannot be supported by experimental evidence at this stage.

Patent applications are normally filed in a wide range of countries representing the major market for agrochemicals or territories within which agrochemical manufacture is likely to take place for export.

The stringency of patent examination within patent offices varies between countries. Generally the national patent examiners will compare the claims made by the applicant with those previously made by other companies in similar areas of chemistry and with the existing literature.

Claims which conflict with prior claims of other inventors, or which are not considered to be a significant advance over knowledge reported in the literature, will not be allowed. Similarly, claims for a breadth of chemical structural types which are not adequately supported by listed, made and tested examples will not be allowed.

Considerable negotiation with each national patent examiner may be needed before the final specification of the patent application is agreed. Once agreement has been reached the patent will be published in official gazettes. There is, usually, a period allowed for objections by other parties before, finally, the patent is granted.

The patent can then be used as a basis for legal action against any other party exploiting the invention before the date of patent expiry.

The general length of life for an agrochemical patent is 20 years. In some territories life is measured from date of first 'priority' application. In others it runs from the date on which the patent was granted.

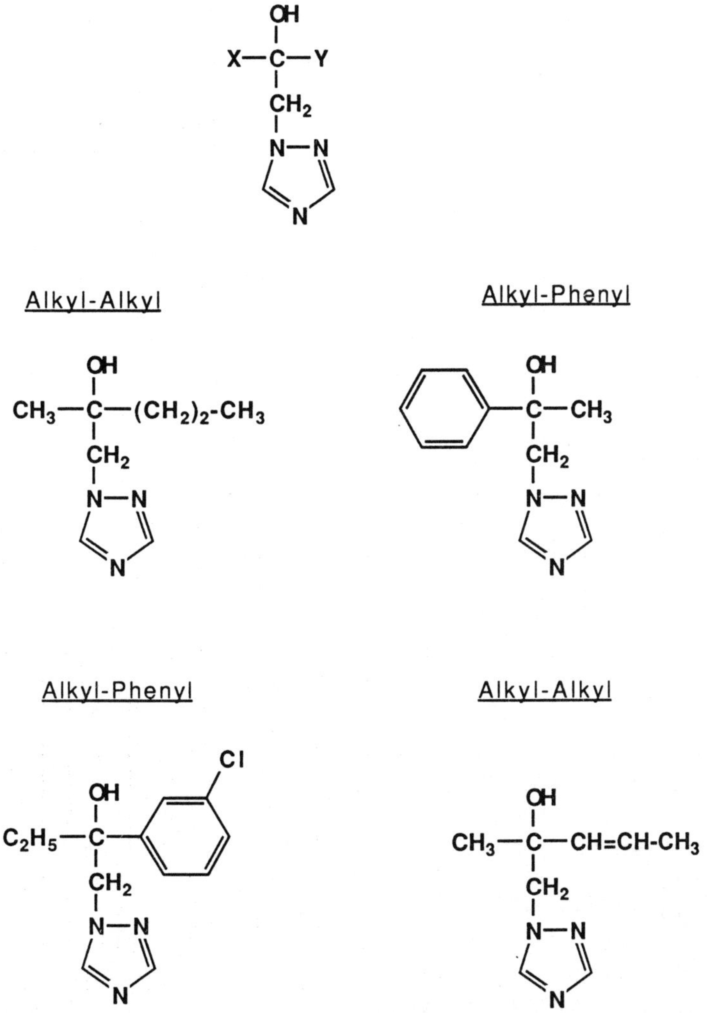

Fig. 20 — Examples of structural analogues within a defined group of triazoles.

Because several chemical companies are often working in the same broad areas of chemistry in their search for good new compounds, dates of patent application can make the difference between commercial success and failure. There are many instances of companies failing to be able to exploit good new inventions because they filed their patents only days or weeks after a competitor working in the same area of chemistry.

In the USA it is possible to claim priority from the date of 'conception', if this can be substantiated by laboratory notebook evidence and appropriate affidavits.

Although patents are essential for protection of inventions, the first publications in novel areas of chemistry do provide valuable sources of ideas for competitors who can then divert effort to synthesize molecules close to, but not embraced by the area covered by the new patent. This immediate following of competitive patent leads has provided sources of valuable inventions.

THE SCREENING PROCESS

Past experience within the agricultural chemicals industry has indicated that only one out of over 10 000 molecules initially tested ever emerges as a practically useful product.

A rigorous selection or screening process is therefore necessary to identify those with practical utility as efficiently and cheaply as possible and to ensure that only those with a high probability of commercial success enter the costly full development process.

An efficient screening process needs to:

— allow a very high throughput of new molecules per year;
— have a low cost per molecule tested;
— have a high power of discrimination between good candidate chemicals and those unlikely to be useful;
— require very small quantities of chemical at the initial stage.

This requires:

— A clear definition of the target biological characteristics of the desired end product;
— A series of very simple tests which give reliable indications of whether a molecule possesses those desired characteristics or not. An excessive number of false positives would lead to the subsequent commitment of resources in error. An excessive number of false negatives would lead to a failure to recognize and progress potentially worthwhile products.

Conventional screening processes to select molecules are constructed as a 'cascade' of tests, each test in the sequence being more complex and discriminatory than the previous one and letting through fewer candidates for the next stage.

Primary screens

The first stage of screening is to determine whether a candidate molecule possesses any activity against target organisms, crops, fungi, insects or weeds. The available quantity of each novel molecule will be very limited at this stage, usually in the order of milligrams, and the testing procedure must be designed accordingly.

At this stage for example, fungicide tests may comprise a number of representative fungal species grown *in vitro* on agar in glass petri dishes. A second stage of primary screening for compounds with some apparent activity may comprise a range of major representative diseases grown *in vivo* on small plants in small pots in a glasshouse (Plate 11). Candidate compounds would be tested at a pre-determined range of concentrations to determine the 'break point', the lowest rate which gives appreciable disease control.

(a)

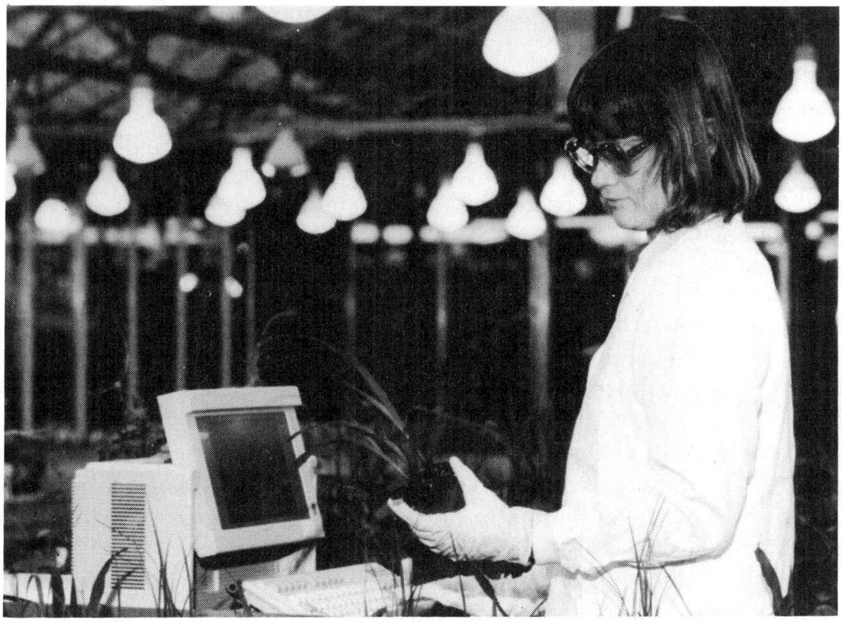

(b)

Plate 11 — Testing new compounds in a glasshouse. (a) Seed trays containing a range of plant species used for testing chemicals for pre- and post-emergence weed control. (b) Assessing young wheat plants for fungal disease control.

Those molecules which show good activity serve as leads for further chemical analogue synthesis and become candidates for more detailed characterization in the next stage of glasshouse screening.

Secondary screening

The next stage of glasshouse screening attempts to characterize biological activity sufficiently well to enable a selection to be made of those compounds worthy of being taken into field screens.

This involves testing against a range of major diseases, weeds, or insects and for absence of phytotoxicity on major target crops. The spectrum of targets controlled and the range of crops on which the molecule appears to show no phytotoxicity, will indicate the commercial targets for which it could be a candidate.

Glasshouse tests can also give some guide to the ways in which the molecule could possibly be used. For example, herbicide screens will indicate whether a compound has potential utility for application to soil to prevent germinating weeds establishing, or for application to established seedling weeds to kill them, or whether it has the potency to control large established weeds. Fungicide secondary screens will indicate whether a compound will eradicate existing disease infections or protect against infections which occur after application. They will also indicate to what extent the chemical will move in plants to control disease at points away from the site of chemical application. These pieces of information give guidance on how to test in later stages for optimum performance and also show how the compound would need to be used in practice if it were to be developed. Advantages or disadvantages compared with current practice would affect decisions to progress.

Although considerable advances in knowledge can be made relatively cheaply, quickly and effectively in glasshouse tests, these are inevitably only imperfect models of field reality and, as a result, leave considerable uncertainty as to the likely field performance under practical conditions.

The final stage of screening, therefore, takes place under field conditions.

Field screening

Each of the major agrochemical companies has field screening stations spanning the world's major crops and climatic zones. At these research stations, the biological efficacy of candidate molecules is characterized under field conditions where severe pressures from target weeds, insects or diseases can be ensured by artificial inoculation or by control of environmental conditions. Field experimental work will concentrate on those targets indicated from the preceding glasshouse work and will attempt to produce a good initial characterization in comparison with existing commercial standards.

By the end of the field screening process the researcher would expect to have defined a broad range within which the likely optimum rate of application would lie ($c \pm 50\%$), the rough timing and methods of use and the major crops and pests on which further work should be concentrated.

FIELD EVALUATION

A proportion of the compounds field-screened will perform sufficiently well to be taken into larger scale field evaluation. In evaluation, substances will be tested on

Plate 12 — Field screening trial plots for weed control in rice.

larger scale field plots. For conventional agrochemicals, about 1 compound in 2000 initially tested reaches this stage and that compound has about a 1 in 5 chance of becoming an acceptable product. It is the first truly realistic representation of the conditions under which the compound would need to perform in practice, if it were to become a product, and provides the last opportunity for a very critical performance appraisal before selection for development and the commitment to a $30-60m programme of work.

During the evaluation phase, work will also be carried out to determine the optimum formulation type, including mixtures with co-pesticides, and to identify, in a preliminary way, the major technological and environmental characteristics of putative products.

During the field evaluation phase a number of preliminary safety studies would also be conducted (see Chapter 5) which would also have a major effect on the decision on whether to progress the compounds into development.

DEVELOPMENT

The development stage addresses the detailed aspects of a potential product:

— biological efficacy,
— safety to users, consumers and the environment,
— production,
— marketability,

Plate 13 — Field development trial plots assessing establishment of barley seedlings treated with fungicidal seed dressings.

— financial viability.

This section discusses work to establish biological performance. The other aspects are considered separately in subsequent chapters.

For a conventional pesticide, biological development will seek to define:

— spectrum of pests, weeds or diseases controlled;
— duration of control obtained following treatment and ability to eradicate established target organisms;
— detailed dose response for optimum efficiency and crop safety;
— optimum timing of application in relation to stage of crop and pest infestation.

By the end of development, precise information must be available to allow:

— clear definition of the optimum rate of application per hectare to achieve the best balance between effectiveness and safety to the crop and the environment;
— clear definition of the optimum timing of application in relation to the stage of crop development and the life cycle and epidemiology of the pest;
— knowledge of the influence of crop variety and environmental variation on performance and crop safety;
— knowledge of the interactions between the newly developed product and others with which it may be mixed in a farmer's spray tank;
— assessment of the possible influence of product use on the development of tolerance or resistance by the target organism.

This information is obtained from a series of large scale, statistically analysed field trials, each planned to answer a specific set of questions under a particular set of geographical and climatic conditions, typical of the major target markets worldwide.

The body of data needed to provide all the information which has just been listed for a product has to be built up piece by piece. For example, to define the optimum rate of application against each target pest, the first step is to conduct a series of field trials which test a range of rates of application on sites where the particular pest is known to be present at high levels or where infestation can be induced artificially. For many fungal diseases, for example, high disease pressures can be ensured by artificial inoculation of the trial crop, using misting equipment to maintain an environment conducive to disease spread or by planting particularly sensitive varieties. From the measurements of disease levels on each plot in the trial it is possible to understand the relationship between biological response and rate of chemical applied per unit area. This is exemplified in Fig. 21 which shows the typical layout of a replicated field trial, example results and the resulting dose-response curve. Similar data will be gathered in at least seven to ten trials for each pest in each agronomic area and a composite picture of dose-response will be assembled across trials which will give some indication of how robust the treatments are. The commercially recommended use rate against the example pest in Fig. 21 would be chosen partly on these data, where the 250 g/ha rate gives a good level of control comparable to the competitive standard. A high rate could be chosen to give technical excellence, or it may be decided that anything over 90% control is commercially satisfactory and a lower application rate may be chosen for economic or environmental reasons.

Assessment made at a number of different times after treatment in the trial would give data on the speed and duration of control obtainable at each rate.

Other trials, possibly using one preferred application rate, testing application at different stages of crop growth or of the epidemiology of the pest, would show how critical the timing of application is for the particular compound. A programme of treatments may be needed at prescribed intervals or growth stages.

Once an application rate and application timing (or range of timings) have been defined, it is necessary to check the reliability of that 'recommendation' in widespread trials covering a wider range of crop varieties and geographical locations. It is also necessary to use the proposed product in mixtures with a range of other products which could be used on the crop at the same time to check that there are no adverse interactions. These interactions may give rise to problems in the spray tank (through incompatibility of formulation) or to increased risks of crop damage, or result in reduced effectiveness of either component.

To carry out these programmes requires well-trained experimental teams in all geographical areas where a product is to be developed and used. For a product to be developed internationally, the chemical company must either have field trials staff distributed throughout the countries in which it seeks to sell or else have established the necessary business relationships with other organizations which have such facilities. This is discussed further in Chapter 9.

During the widescale testing procedure, practical problems will inevitably be encountered which can be solved only by more basic research. Problems may be encountered with the formulation (see Chapter 6), and changes may have to be made

Randomized Field Plots

Plots	Replicates I	II	III	IV
1	1	5	2	3
2	4	2	1	4
3	2	3	4	1
4	5	1	3	5
5	3	4	5	2

Results

Treatment	Rate g/ha	% of area of wheat flag leaf infected with brown rust Rep I	Rep II	Rep III	Rep IV	Mean	% Control
1. Compound A	125	5	6	4	3	4.5	89
2. Compound A	250	2	3	2	1	2.0	95
3. Compound A	500	0	1	1	0	0.5	99
4. Standard	375	4	2	3	1	2.5	94
5. Untreated	—	43	39	45	40	41.8	—

Dose–response curve

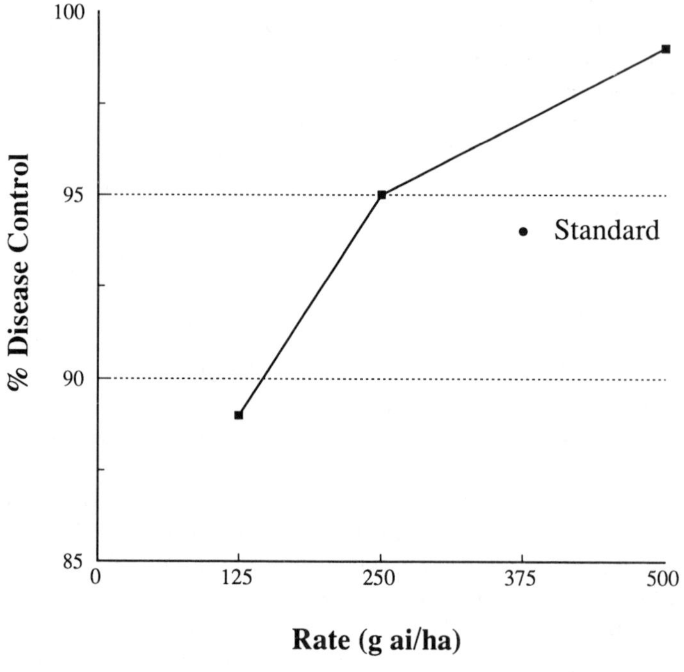

Fig. 21 — Example of a typical field trial.

to the surfactant system, for example, to improve performance, reduce phytotoxicity or improve compatibility. Changes may be needed to overcome practical handling problems in particular types of spray equipment.

Often, in a programme of field trials, there are a few tests in which the product performs atypically. A study of the peculiar conditions occurring in that trial is often necessary to gain a further insight into the product's characteristics. This may, in turn, give rise to yet more research to minimize the risk of failure later under commercial conditions. The problem may be overcome through a slight modification to the product or by writing specific instructions in the product use label to avoid these situations, for example label warnings against using the product on particular soil types or crop varieties.

In a number of countries, the results of these trials must be submitted to government regulatory authorities as part of a dossier of data to seek registration and permits to sell. In some countries (e.g. West Germany and Japan), tests must be performed by recognized officials on government and other approved testing stations. There is often a requirement that tests should be carried out over at least two seasons to give more robust evidence of consistency of performance.

The findings of all these field trials are finally distilled into a set of product use information supplied to farmers in the form of a product use label attached to the sales pack and in supporting product use literature. An example of a product use label is given in Appendix 2.

Development of a compound does not cease once it is marketed. As a chemical becomes established in practical use, further mixtures, crops or methods of use may become established. New mixtures, formulations or uses for compounds can still be in development twenty or thirty years after the first sales were made. Each new application will require further programmes of supporting work so that approval can be obtained from government registration authorities. Once this has been obtained the new usage or application can be added to the product use label and literature.

5

Product safety

THE EARLY CONCERNS

The developments in the use of crop protection chemicals have been paralleled by an increasing awareness of, and concern for, the consequences of widespread chemical application with regard to the spray operator, the food consumer and the environment. The first concerns were for direct human consequences.

In the earliest days of the use of Bordeaux mixture, in 1885 and 1886, there were public outcries about the risk to consumers posed by the widespread use of copper sprays on grapes and the residues that could result. At that time, Gayon tried to set these fears at rest and put the residue levels in perspective. He showed that wine made from grapes sprayed with Bordeaux mixture contained only 0.33 ppm ($\frac{1}{3}$ mg/l) of copper, less than the concentration in water raised by copper pumps and much less than that found in common foods such as meat and chocolate. Complementing this, Millardet demonstrated the uniquely high level of toxicity of copper to fungal spores, 0.2–0.3 ppm in water giving complete inhibition of *Peronospora* spore germination (Large, 1940).

The next outcry came in the early 1890s in the English press expressing concern about the hazards to English consumers of American apples treated with arsenical insecticides. Riley of the United States Department of Agriculture (USDA) countered this with estimates that one would need to eat 38 barrels of apples, including the core, to absorb dangerous levels of arsenic. These types of calculation foreshadowed the more sophisticated risk assessments which play a major part in the regulation of chemical use today.

Many of the early arsenical insecticides did, however, pose a significant hazard to spray operators and deaths occurred as a result of their misuse. During the early part of this century, awareness grew of the need to treat agrochemicals with a certain degree of care.

THE GROWTH OF AWARENESS

All chemicals, whether natural or synthetic, have the potential to interact with the biochemical processes of life (and some of the most lethal are naturally occurring substances). The rapid expansion in the use of agricultural chemicals as an integral part of crop production has, very properly, led to an increasing awareness of the need to understand the long term consequences for users, food consumers and the environment. The developments in analytical chemistry and understanding of biochemistry which have contributed to the discovery and advance of agrochemicals have enabled far more searching questions to be asked, and answered, about the fate of those products in mammalian systems and the environment.

Since the mid-1960s, spurred on by publications such as *Silent Spring* in 1963, political pressure has ensured that not only are searching questions asked but also that they are answered to the satisfaction of specialist government authorities before a product can be registered for sale (Carson, 1963).

The development of international knowledge about the behaviour of agrochemicals in the environment tended, during the 1960s and early 1970s, to lead to a desire by government registration bodies to raise further questions aimed at an even greater understanding. Some indication of the growth in the range of safety topics to be studied for each new product is illustrated by Table 12.

The detailed study of the toxicological and environmental fate of a new agrochemical forms a major part of the development phase and can contribute 40–50% of its costs.

As a result of these studies the fate of agrochemicals in the environment is probably better understood than that of any other class of substances.

Each aspect of safety package of studies required to demonstrate safety is discussed in detail in the following section.

HUMAN SAFETY

Toxicology

The first concerns for human safety arising from pesticide use arose in the late nineteenth century because of the acute toxicty of the early insecticides, notably the arsenates. The toxicity problems were clearly manifest in the sickness and death of spray operators.

As chemical usage and scientific knowledge have grown, a wider range of sub-acute and chronic effects have achieved increased public awareness, notably any predisposition to cancer and any possible risks to unborn children as a result of maternal exposure.

For both operator and consumer safety assessments it is now necessary to investigate, before registration of a new pesticide product, the likelihood of exposure to that chemical causing any effects on mammalian metabolism, histopathology or reproduction or on the offspring of individuals exposed to the chemical in any way.

Studies required include:

(a) *For operator safety*

- Acute toxicology intrinsically (measured by injection) and as a result of oral ingestion, inhalation or dermal contact.
- Irritancy to skin and eyes.

Table 12 — Trends in registration study requirements (general trends; detailed requirements differ between countries)

	1950	1960	1970	1980[c]
Toxicology	Acute toxicity 30–90 day rat	Acute toxicity 90-day rat 90-day dog 2-year rat 1-year rat	Acute toxicity Irritation Skin sensitization 90-day rat 90-day dog 2-year rat 2-year mouse Reproduction (2–3 gen. rat) Rodent teratology Fish, shellfish, etc. Birds	Acute toxicity Irritation 90-day rat 90-day dog 1-year dog 2-year rat 2-year mouse 2–3 gen. rat reprod'n Rat & rabbit teratology 3–5 mutagenicity assays Fish, shellfish, etc. Birds Soil organisms Beneficial species screen
Metabolism		Animal	Rodent and/or dog Target plant	Rodent Range of target plant species Rotated crops Cattle, pigs, poultry
Analytical	Food crops 1 ppm[a]	Food crops 0.1 ppm[b] Meat 0.1 ppm Milk 0.1 ppm	Food crops 0.01–0.05 ppm Meat 0.1 ppm Milk 0.005–0.03 ppm	Food crops 0.01 ppm Water 0.1 ppb (EEC standard)
Ecology			Environmental stability Degradation in soil in lab. Movement in soil in lab. Spectrum of affected species Accumulation	Environmental stability Tiered studies of degradation & movement in lab. and field. Field monitoring studies on accumulation & ecological effects. Life cycle studies

[a] Pesticides only.
[b] Pesticides plus toxic metabolites.
[c] Many protocols have been revised and been made even more thorough since 1980.

— The effect of repeated exposure which may cause increased sensitivity to the chemical.
— For products likely to cause some adverse effects, spray operator monitoring studies to assess practical exposure and its consequences, e.g. the effect of organophosphorus insecticides on the blood cholinesterase levels of spray operators.

(b) *For operator and consumer safety*

- — Effects of long term daily ingestion of the compound on blood chemistry, histology of major organs and risk of enhanced tumour formation.
- — Effects on overall health and behaviour.
- — Rates of accumulation and/or excretion of the pesticide molecule or any of its metabolic products. For products that do accumulate in body tissue, studies are required to show the tissues in which that accumulation takes place and the rate at which those residues can be eliminated again once exposure ceases.
- — Elucidation of the metabolic pathways of the pesticide and its metabolic products within mammalian systems. For example the metabolic pathway for the insecticide permethrin in rats is illustrated by Fig. 22.
- — Risks of effects to offspring of exposed male or female parents. Teratology studies are required to show any effects of maternal exposure during pregnancy on foetal growth rates and development and on reproductive performance over at least two generations.

Because of the ethical and economic objections to conducting a comprehensive toxicological programme on chemicals which are likely to prove to be hazardous, and therefore unacceptable, candidate chemicals are first subjected to a number of indicative tests. Detailed studies are carried out on chemicals which pass the indicative tests and form part of the evaluation phase preceding full development.

Many of these studies have to be performed using laboratory animals in ways specified in detail by government regulation. Others can be conducted using simple organisms, such as bacteria, or cell cultures to provide indicative information, e.g. the Ames test as a preliminary screen for potential carcinogens (Ames, 1975). Chemicals are tested against cultures of bacteria, e.g. *Escherichia coli* or *Salmonella typhimurium*, for their ability to cause genetic mutation. Tendency to cause mutation suggests higher risk of carcinogenic potential, while freedom from mutagenic effect indicates lower long term risk. High risk candidates can be screened out before longer term mammalian tests. The detailed relationship between mutation and cancer is, however, not well-understood.

Tissue cultures and enzyme systems can also be used to test candidate compounds for risk in certain well-characterized areas of toxicology. Animal studies are however, vital to establish the subtle and overt long term effects of exposure and the potential risk of, for example, increased carcinomas. Taken together, the long term studies enable assessments to be made of what health effects, if any, are likely to arise from long term repeated exposure and, just as importantly, at what level no effects occur. The 'no effect' level is a major criterion in assessing safety when taken together with measurements of likely exposure risk.

Often the effects observed will prove to be very slight and discernible only through very detailed studies of haematology, clinical chemistry or histopathology.

Long term feeding studies monitor in detail any changes in the physical or chemical composition of the blood during the progress of the study and any behavioural or weight abnormalities. On post-mortem, sections of all major organs are taken, mounted and stained and examined by skilled pathologists for abnormalities.

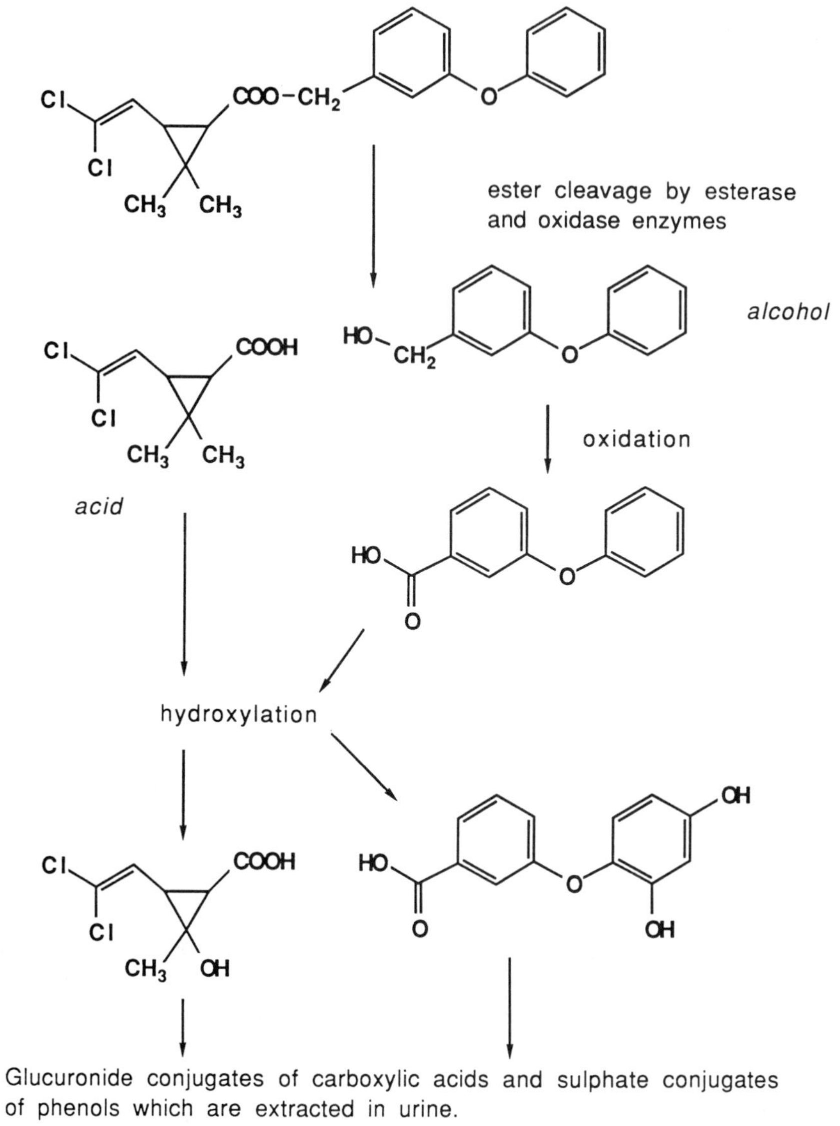

Fig. 22 — Metabolism of permethrin in rat.

The statistical and toxicological significance of these changes are then assessed in relation to humans and suitable safety factors applied to the no-effect level, which may be 100- or 1000-fold depending on the nature of the effect. The setting of safety factors is discussed later in relation to overall risk assessment.

Acute studies are done to indicate the general nature and degree of toxicity of the pesticide and a short programme of research toxicology is conducted to indicate major sub-acute effects and target organs. Short non-animal tests are also used to assess mutagenic and carcinogenic potential. Many agents can cause alterations to the genetic characteristics of a cell, common examples being ultra-violet light or X-rays.

The consequences may be changes in reproductive fitness or increased risk of cancers. Effects on reproduction can be in conception rates, litter sizes, or in the health and normality of the offspring, studied in longer term mammalian tests.

Since the toxicology tests conducted reflect the toxicological properties of the agrochemical molecule *and* any concomitant impurities it is important that any impurities present in the material tested are known and evaluated. Ideally the material tested should be representative of the material as eventually produced in commercial manufacture (see Chapter 6). In practice, however, as the development of manufacturing routes is a long and complex process, decisions on the preparation of toxicology testing samples have to be taken before development of the manufacturing process is complete. If there should be high levels of impurities which are relatively more toxic than the product under test, there could be serious adverse effects on the whole testing programme. For these reasons the manufacturing process for agrochemicals aims to provide very high levels of purity, which can have serious implications for production costs.

Very refined analytical methods have to be developed for each new agrochemical to measure levels of the parent molecule, its impurities and its metabolites, often down to levels of 1 part per thousand million (a concentration comparable to half a teaspoon of sugar dissolved in an Olympic size swimming pool or 1 second in 32 years).

Chemical intake in crop residues

Another important part of the consumer hazard equation is the level of chemical residue remaining in a crop when eaten. All hazard assessments use the conservative assumption that for unprocessed crops, residues at the time of consumption will be equal to those at harvest, whereas in practice residues are likely to decay. For processed crops, studies are required of the fate of residue during processing. This will be discussed further later. Chemical residue data are required for both the parent pesticide and its metabolites in plants.

Plant metabolism studies with appropriately radio-labelled chemicals define for appropriate major crop groups the breakdown pathway for the pesticide in the plant, and the sites of deposition of the metabolite residues and parent product. Fig. 23 illustrates the metabolic degradation of the insecticide permethrin in cotton and beans as an example.

The permethrin was sprayed onto the leaves of the cotton and bean plants as it would be in commercial use. The chemical breaks down on the plant and half has decomposed in 7–9 days. The breakdown processes are essentially the same in both plant species. The first decomposition stage, breaking the compound into its alcohol and acid components, is also the same as that shown in rats (Fig. 22). Unlike rats, plants do not readily oxidize the alcohol.

The definition of metabolites is followed by the development of analytical methods for measuring residues which arise under field conditions following use at the proposed rates and timings.

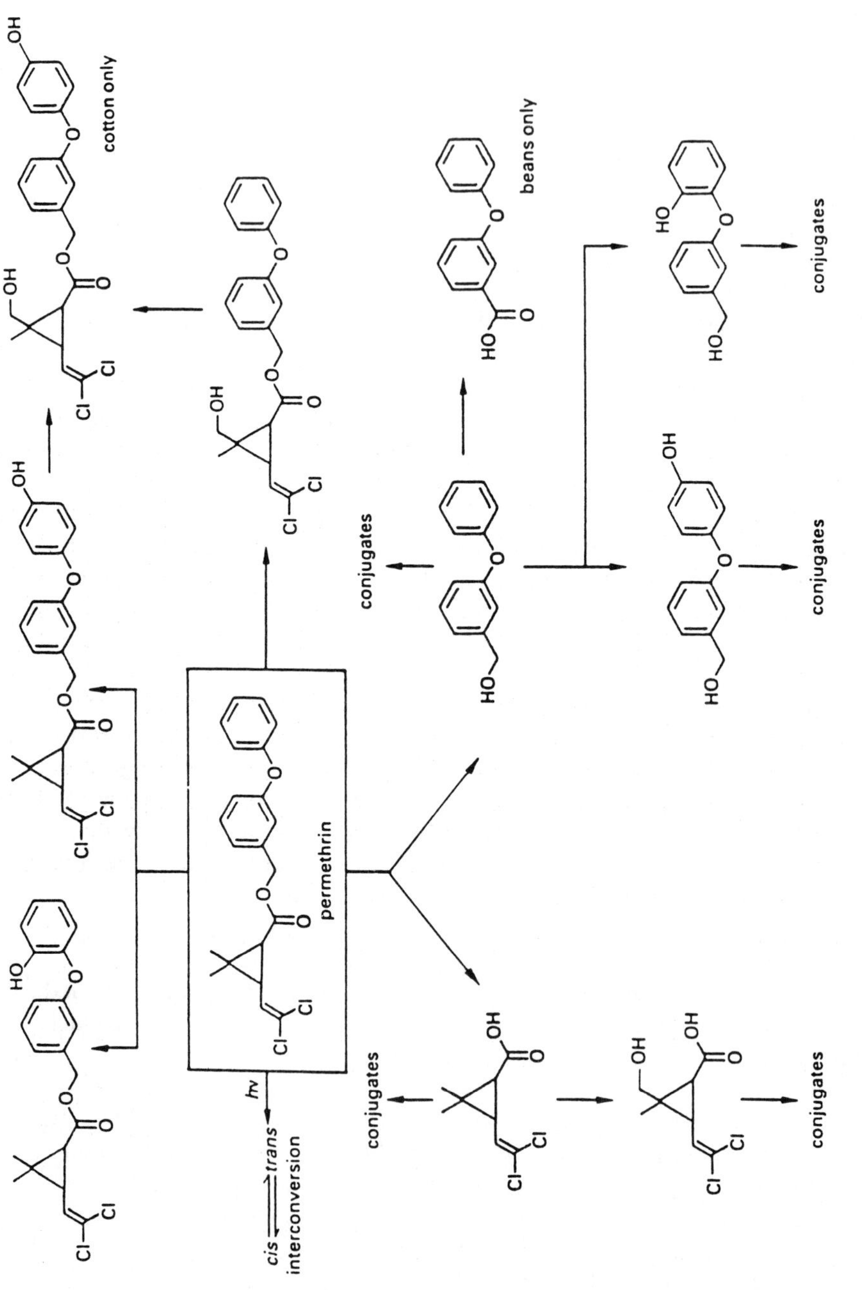

Fig. 23 — Phototransformation and metabolism of permethrin or cotton and bean plants (Leahey, 1985).

In field studies, the chemicals are applied in commercial formulations at recommended usage rates under practical conditions and harvest samples are taken for analysis of residues of parent and major metabolites. Studies are also carried out to measure the rate of degradation of residues in the period between application and harvest, to enable acceptable harvest intervals to be derived.

Crop rotation studies investigate the possibility that residues of an agrochemical (or of its breakdown products) sprayed onto one crop may remain in the soil and be taken up by a following crop grown in the same field in the next season. The chemical is applied under controlled conditions to the 'target' crop, which is then harvested and a range of other crops are planted on the same plots at intervals, grown, harvested and analysed for residues of parent and metabolites.

If a crop is commonly processed before human consumption (e.g., wheat made into bread, grapes made into wine, apples made into juice) residues have to be followed through these process to measure any changes in concentration. If the crop is commonly fed to livestock, then in animal transfer studies, residues must be followed through the animals into meat, milk or eggs. These studies will be carried out both with radio-labelled chemical, to enable detection and identification of any novel metabolic products, and then on a larger scale with 'cold' material to enable quantitative determination of likely residue levels under practical conditions.

Commonly the metabolic products generated within plants are the same as those generated in mammals. When this is so, the toxicology of the plant metabolite is automatically encompassed within the mammalian toxicology programme on the parent compound. When plant metabolism gives rise to metabolites not found within the mammalian system, then a number of government authorities demand special toxicology programmes on those metabolites as well.

From knowledge of the mammalian toxicological 'no effect' levels and by application of large safety factors, set by government regulatory authorities, it is possible to set maximum acceptable daily intake levels for agrochemical residues. Using standard consumption factors for crop or crop grouping it is then possible to calculate maximum residue tolerance levels permissible within that crop. These can then be compared with the actual residue levels observed in field experiments. This may be illustrated by the logic of an exemplary calculation:

— For a hypothetical agrochemical, the no observed effect level (NOEL) in long term feeding trials is 10 mg ai/day/kg body weight of the most sensitive species, e.g. mouse. The 'no effect' level is the feeding rate in the tested dose response *below that* at which *any* effect was seen. For example groups of test animals could be fed at dietary levels of 100, 50, 10 and 1 mg/kg/day. The lowest rate at which an effect was seen may be 50 mg, indicating 10 mg/kg/day to be the NOEL. Effects may be statistically significant changes in histology, body chemistry or animal behaviour between treated and untreated animals. The effect at the lowest effect rate is usually very far from being life-threatening to the species.

— Regulatory authorities then apply a safety factor to this NOEL, usually 100-fold. In some circumstances where the effect is well-studied in human volunteer exposure monitoring and is easily reversible, e.g. lowering of cholinesterase levels following exposure to organophosphorus insecticides, a 10-fold factor may be permissible. In other situations where the effects are potentially more serious, e.g. teratogenicity, or where databases are inconclusive or equivocal, larger factors will be applied, e.g. 1000-fold. Using the 100-fold factor, the assumed acceptable intake level for man is 10/100 mg/kg/day = 0.1 mg/kg/day.

— The weight of the 'average western man' is taken to be 70 kg, giving an acceptable daily intake level (ADI) of 70×0.1 mg/day = 7 mg/day.

— The *maximum* level of residue of the agrochemical in food must be sufficiently low for the ADI, from all sources of potential consumption added together, not to be exceeded.

— For compounds which show any signs of increasing risk of cancer formation (suspected carcinogens), risk assessments often consider total intake over a lifetime rather than daily intake. This is because carcinogenic risk is thought to depend upon cumulative exposure.

Risk assessments may also make special consideration of certain sectors of the population thought to be at increased risk. For example, if a particular agricultural commodity makes up a higher than average proportion of the diet of one sector of the population, e.g. babies, then separate risk assessments may be made for them.

— The *actual* level of residue in any one commodity is measured from practical field experiments. Residue trials are carried out using the maximum rates of application, the maximum number of applications and the minimum intervals between last application and harvest specified by the registered use label. The tolerance (the maximum permissible residue in the commodity) is requested on the basis of the highest level noted in trials. If, for example, the highest observed level was 0.7 ppm, the requested tolerance might be 1 ppm to allow some margin for further variability. In practice, the average residue levels will be much less because:

- usage rate may be less than the maximum permitted,
- numbers of applications will often be less than maximum permitted,
- intervals between last application and harvest will often be longer than the permitted minimum,
- degradation will occur between harvest and consumption (samples for analysis are always deep frozen at harvest and therefore show no degradation),
- only a proportion of the nation's crop will be treated.

Average residue levels may be, for this example, 0.1 ppm.

— Permitted tolerance levels are compared against ADI levels using data on food consumption levels. A number of authorities, for example the World Health Authority and the US Environmental Protection Agency (EPA), maintain tables of dietary contribution for all foodstuffs. If, for example, the agrochemical is used only on fresh apples and the average apple consumption is 50 g/person/day, the dietary exposure from a 1 ppm (1 mg/kg) exposure would be 0.05 mg/person/day.

— Comparing 0.05 mg/person/day with the ADI of 7 mg/day shows an additional 140-fold safety factor.

— The total safety factor for the worked example would be between the *actual* residue level of 0.1 ppm, giving an intake of 0.00007 mg/kg body weight of man/day, and the 'no effect' level on the most sensitive species of 10 mg/kg/day, namely of over 10^5 times.

— The eventual risk assessments will add together residues from all possible sources including any transfers from treated crop to meat, milk and eggs and any concentration in food processing (e.g. juicing fruit).

Particular recommendations for product use are only registerable if the actual residue levels fall below the maximum tolerance levels set.

A number of national governments not only demand mammalian toxicology and crop residue data before an agrochemical can be registered for use in their countries but also demand the information so that they can set import tolerances for foodstuffs treated in other countries.

Countries such as the USA will monitor food imports for pesticide residues and will turn back cargoes of products containing residues not conforming with US tolerances.

For chemicals used on crops commonly traded internationally, such as wheat, manufacturers seek internationally set tolerances from a FAO/WHO joint committee on pesticide residues.

SAFETY TO THE ENVIRONMENT

The enviromental impact of widespread spraying of biologically active molecules has raised an increasing number of questions since the issues were dramatized in *Silent Spring* and the accumulation and environmental consequences of organochlorine insecticides such as DDT became commonly known. There are now comprehensive environmental safeguards built into pesticide registration legislation, and the agrochemical industry often supplements the official requirements with additional studies in areas which may be of particular importance for specific products.

The package of environmental studies required is designed to produce an extremely detailed picture of the fate of the parent compound and its breakdown products in every component of the environment and the consequences of its presence for a representative selection of non-target organisms ranging from soil microflora and microfauna to mammals.

Species considered include:

Mammals	Rat, mouse, dog
Birds	Duck, quail, pheasant, birds of prey (e.g. for rodenticides)
Fish	Trout, carp, bluegill sunfish
Shellfish	Shrimp, oyster
Beneficial insects	Bee
Soil fauna	Earthworms, micro-organisms (including effects of chemical on nitrogen and carbon turnover in soil)
Aquatic micro-organisms	Daphnia

Its chemical decomposition pathways are defined under a wide range of conditions. Metabolism in plants has already been discussed. Breakdown patterns due to hydrolysis (decomposition in water), photolysis (decomposition in strong light) and degradation in soil (by chemical or microbial means) must also be elucidated.

The hydrolytic degradation of a compound is defined in acid, neutral or alkaline water to determine the rate of simple hydrolysis and the nature of the breakdown products.

The photolytic degradation of a compound is studied in water under high UV light intensities. Photolysis is also looked at in pond water and on soil surfaces to determine the possible influence of naturally occurring substances which could participate in photolytic reactions.

The degradation rate and the nature of the breakdown products in soil are defined using radio-labelled compounds in a range of soil types under both aerobic and anaerobic soil conditions. A series of longer term confirmatory studies conducted under practical field conditions are also required by some authorities. For compounds which degrade relatively slowly and rely upon microbial action rather than simple chemical processes for their breakdown, field studies often provide a very much better guide to practical environmental impact in real life than do laboratory studies. Laboratory conditions may enhance or impede the natural rate.

The potential risk of chemical residues in soil moving into ground water systems is assessed through studies of the adsorption and desorption of the chemical to and from soil particles and the leaching of the chemical and its breakdown products in soil.

Leaching experiments are carried out in the laboratory with soil columns in cylindrical containers (Plate 14). Water is passed through a vertical core of soil into which chemical has been incorporated. The leachate is captured in a receptacle at the bottom of the column and is periodically analysed for parent and breakdown products. Horizontal segments of the soil column are analysed at the end of the experiment to measure overall chemical movement.

Some authorities require additional aged leaching studies where soil treated with chemical is incubated for a period of time to generate breakdown products whose leaching properties are then investigated. Again, confirmatory field soil residue analysis may be required to define movement potential under practical conditions. These may be carried out in lysimeters or by sampling groundwater.

For some products in widespread use deemed to have leaching potential, the US Environmental Protection Agency has demanded ground water monitoring studies in order to check contamination risks. Increasing political concern about the risks of residues in drinking water have forced more expensive and complex field studies of leaching, coupled with demands for analytical methods to measure concentrations down to 0.1 parts per thousand million.

Influence on non-target animals

The range of non-target organisms is immense, as are the interactions between them in food chains. Some of this complexity is illustrated by Plates 16 and 17. Environmental safety programmes seek to assess the level of toxicity of a new pesticide to a representative selection of these organisms and to give some indication of whether there is a risk of accumulation or secondary effects through the food chain.

Risk of toxicity to mammals is well understood from the programme of studies conducted to assess likely risks to human safety. Bird studies investigate acute toxicity and reproductive effects. Acute studies are conducted on bees and beneficial insects (including soil insects). Studies of the risks to soil organisms also consider

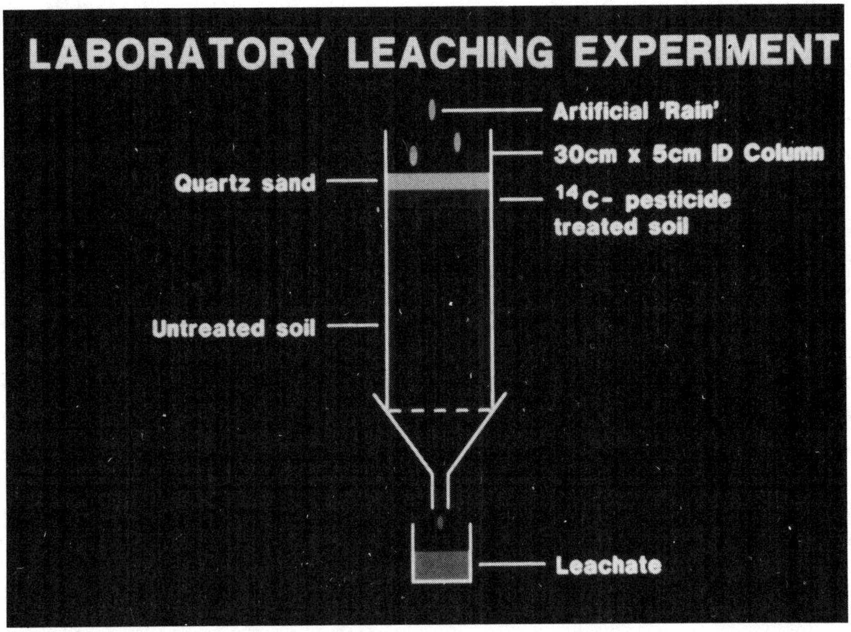

Plate 14 — Soil column to measure leading of pesticides and metabolites under controlled laboratory conditions.

acute and field chronic effects on earthworms and effects on soil microflora in terms of carbon and nitrogen turnover.

The implications of an agrochemical reaching water, through spray drift, leaching, spillage or factory effluent are studied through toxicity tests on a range of aquatic species. For example, the US Environmental Protection Agency protocols cover acute toxicity to fish, oysters, shrimps and Daphnia, reproductive effects on fish and shrimps and accumulation and excretion in fish. For some chemicals, complex pond studies are required to model behaviour in an aquatic ecosystem.

RISK ASSESSMENTS

Agricultural chemicals are subjected to an investigation process which is considerably more searching than that applied to most other aspects of man's environment. Every chemical is different and the results of one set of studies trigger further questions from registration authorities which will then require a further series of studies. All aspects of life carry some degree of risk and it is generally accepted that a risk-free environment is an impossibility. For any particular agrochemical, judgement is required to establish the correct balance between risk and reward. As discussed in Chapters 1 and 2, agrochemicals are now an essential and integral part of crop production systems. Without them, losses of yield due to pests, weeds and diseases would cut world food production drastically; estimates of the losses which

Plate 15 — Lysimeters for measuring chemical movement in soil under field conditions.

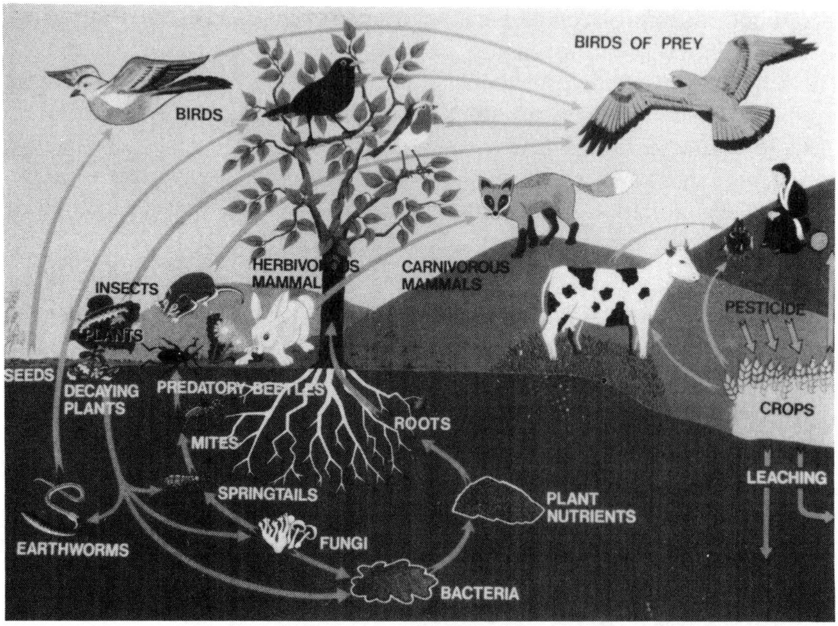

Plate 16 — Terrestrial ecosystem showing possible entry of chemicals in the food chain.

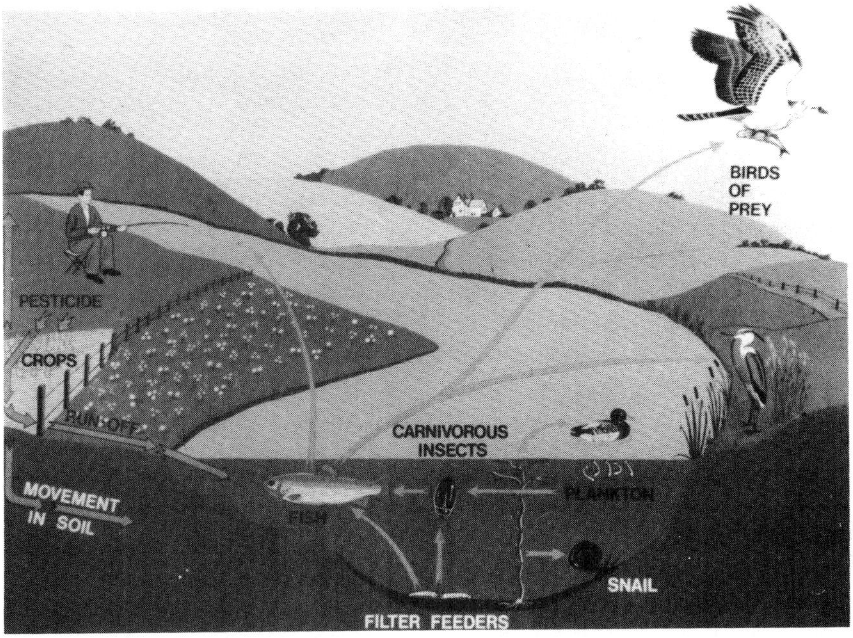

Plate 17 — Aquatic ecosystem showing possible entry of chemicals to the aquatic environment through direct means of run-off, or by leaching through the soil.

would result are very hard to make but have been assessed as in the order of 30% of production and food prices would rise by 50 to 70% as a result. The extent to which any one particular pesticide contributes to the productivity of agriculture in relation to the risks associated with its use is often very difficult to quantify. The US Environmental Protection Agency (EPA) has attempted to make the risk/benefit analysis explicit through its current review procedures.

The elements of EPA risk assessment are:

- Hazard identification (determining the existence and nature of adverse effects associated with exposure to a substance).
- Dose-response evaluation (the relationship between the dose administered and the incidence of the adverse effect).
- Exposure assessment (determining the human exposure to the substance).
- Risk characterization (combining the results of exposure assessment and dose-response assessment to quantify risk).

Hazard identification is a qualitative process to assess whether exposure to a pesticide leads to an increase in occurrence of any particular adverse effect. It draws on information from the array of scientific studies available.

Dose-response evaluation is a quantitative process which seeks to define the lowest dose level at which any adverse effect is caused and to measure whether the effect is altered by dose rate. A number of mathematical models have been developed to relate high-dose experimental results to low-exposure levels expected in practical use.

Exposure assessment seeks to quantify the magnitude, frequency, duration and route of exposure. There are formal guidelines for human and non-human exposure assessments.

Risk characterization takes account of the strengths and weaknesses in each component of the assessment, major assumptions, scientific judgements and, as far as possible, estimates of uncertainties.

The outcome of this assessment is then reviewed in relation to the benefits to be gained from the pesticide's use, with the objective of obtaining the maximum benefits coincident with the minimum risk.

EPA uses several different measures for risk management:

— Cost/benefit analysis (weighing costs of pest control against monetary reward where all factors in a decision to use a pesticide can be measured financially).
— Risk/benefit analysis (weighing economic benefits against assessments of associated risks to health and the environment).
— Cost-effectiveness analysis which accepts an effect as desirable and seeks the lowest cost of achieving that goal.

Balanced views of risk and reward are an essential part of future decision making about the development and use of crop protection chemicals.

6

Production

FROM NOVELTY TO UTILITY

Before a new agrochemical can have any practical, economic use, a cost-effective process for manufacturing the active ingredient must be developed. The chemical must also be formulated into a usable product which has all the physical properties required for easy handling, optimum biological performance and good storage life under a wide range of storage conditions.

This chapter aims to give a broad understanding of the activities which are involved in the development of manufacturing and formulation processes. It outlines the broad objectives which need to be met if an agrochemical is to be produced economically on a commercial scale and sold to the farmer in a package designed to be safe, effective and convenient.

DEVELOPMENT OF MANUFACTURING PROCESSES

When new molecules are made for the first time, the synthetic processes used in laboratory apparatus are often at the boundaries of knowledge of synthetic chemistry. The reagents used and the reaction conditions often bear no relationship to currently operated commercial, large scale chemical production.

The very early synthesis of material for the first screening tests (see Chapter 4) will be only in milligram quantities. Laboratory synthesis will be scaled up to gram and then kilogram quantities for early field tests, but the processes used may still be quite unsuitable for large scale production.

The target for the process chemist is to move from this situation to one in which a commercially practical manufacture is possible on a 100- or 1000-t scale. The contrast between the apparatus used for initial synthesis and that for commercial production is illustrated by Plates 18 and 19.

The establishment of an economic commercial manufacturing process requires:

— the invention of an optimum route;
— process research to optimize the reaction conditions for each step of that route;

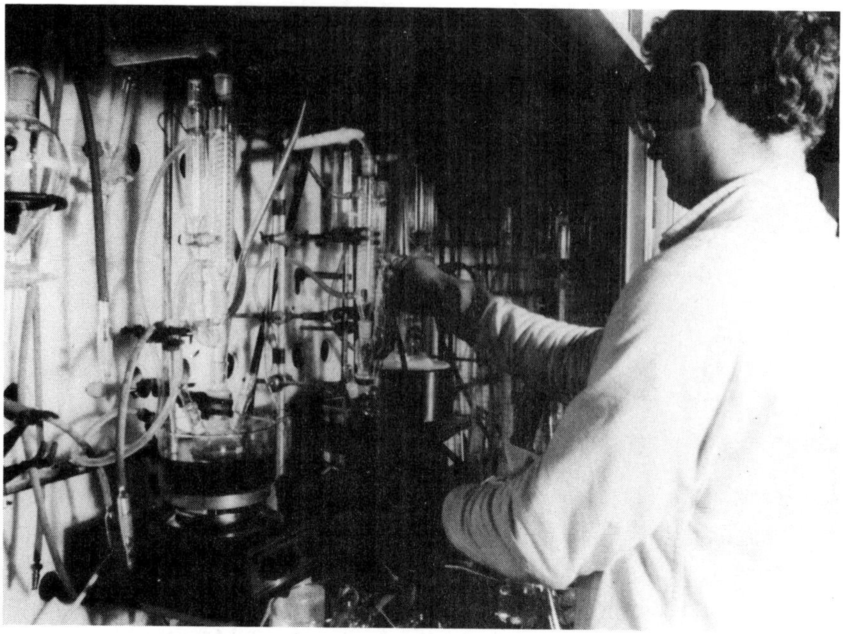

Plate 18 — Laboratory apparatus for chemical synthesis.

Plate 19 — Commercial chemical manufacturing plant.

- — process research to extract the desired product from the end of one step ready for the next step;
- — chemical engineering development to design and enable the construction of a manufacturing plant in which each step of the process can be performed under the optimum conditions. It must be possible for the product of each stage of manufacture to be isolated and passed on to the next stage.

Each of these phases of process development is illustrated in more detail in the following sections.

ROUTE INVENTION

At this point the chemical structure of the desired product is known and the challenge is to devise a synthesis in which the molecule can be built up by a sequence of chemical modifications in a safe, efficient and cost-effective procedure. While the target product is known, there may be a wide range of potential starting materials and alternative synthetic routes. In principle the process could start with naturally occurring raw materials such as oil, coal or rock sulphur, but then the route would be very long and expensive in investment in manufacturing equipment. As the chemical industry produces a vast range of more sophisticated products, this is seldom necessary.

The ideal manufacturing route starts with cheap, readily available, easy to handle raw materials and joins them together in the minimum number of steps under conditions which are easy to provide safely in cheap and easily made reaction vessels. The process should give the minimum of by-products, which should themselves be innocuous. In practice, it is often difficult to achieve this and significant advances in chemical technology may be required to produce a satisfactory process.

The first stage of the invention process is usually a 'paper chemistry brainstorm' to set out every conceivable route to a new molecule, using the best available knowledge of which chemical reactions are possible and which are not. Some of this knowledge will come from empirical precedent, reactions having been demonstrated in the chemical literature or in the experience of the chemists involved. Some will suggest themselves from basic theory of thermodynamics and reaction kinetics. Usually, ideas will not be constrained by whether each proposed step has actually been *proven* in practice.

Experienced process chemists then peruse the range of alternatives and arrive at a short list of options which they believe will give the greatest chance of economic success.

Work will then start to prove these routes; that is to demonstrate whether under laboratory conditions each proposed step of the process will actually work. Some reaction steps in the proposed route may be known from precedent and can be proven relatively easily. Others may be postulated to be possible in theory but never previously attempted in practice. Skill, inspiration and knowledge of theoretical chemistry are necessary to investigate these steps. The chemists must be prepared to challenge conventional wisdom if an otherwise elegant route appears constrained by one step, or if a particular raw material appears to be difficult to obtain.

The practical investigation of the short list of route options should result in a firm preference being established for one or two options or for one main route with one or two slight variants. The selection will reflect likely cost and success, and also take into account the likely by-products and means of disposal of waste products with minimal environmental impact.

The next stage is to carry out process research on each step in turn to define the optimum reaction conditions and the method of isolation and degree of purity needed for the intermediate product from each step before that intermediate can be passed on to the next reaction.

Experimental work to optimize a reaction step involves tests to define:

— optimum solvents in relation to the reactants and the reaction products,
— concentrations of reagents in the solvent,
— reaction temperature,
— reaction pressure,
— duration of reaction time,
— influence of catalysts,
— effect of relative proportions of reagents at different stages of the reactions,
— stirring, mixing and heat transfer aids required to ensure maximum reaction efficiency,
— methods for minimizing or avoiding by-product formation,
— potential hazards and ways of containing them,
— corrosion problems and suitable materials of construction.

Many chemical reactions can be highly sensitive to relatively small deviations from optimum conditions. This is illustrated by two examples, one the effect of solvent used, the other showing effect of temperature.

The solvent used as the reaction medium can have a major effect on the outcome of a chemical reaction. In the manufacture of triazole fungicides, the triazole part of the molecule (ii) can be joined to the oxirane structure (i) in a base-catalysed reaction.

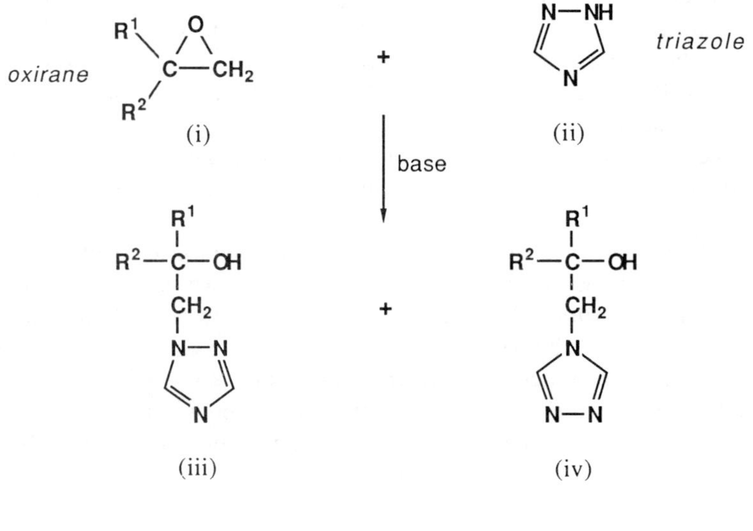

The reaction can produce two isomers (iii) and (iv). The relative proportions of each product depends upon the solvent used. The effective fungicide is the form with the asymmetrical triazole (iii) and its proportion must be as high as possible. Use of dimethylformamide as a solvent gives a larger proportion of (iii) than, for example, propanol.

Temperature can affect the speed of a reaction and also the nature of its products. This is illustrated by another example from triazole fungicide manufacture.

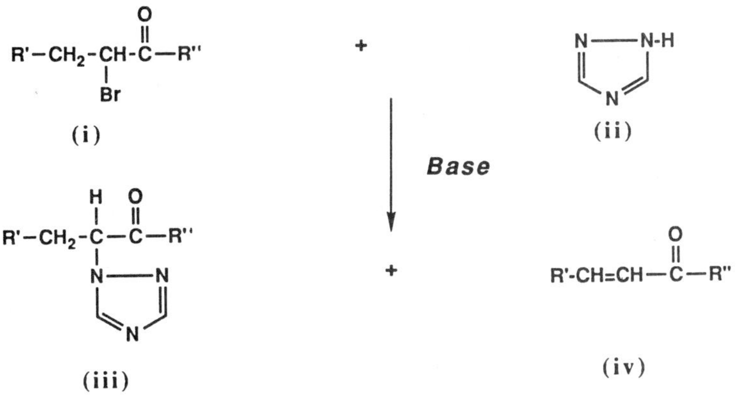

This reaction to replace bromine with triazole produces the desired triazole product (iii) and a by-product (iv). The relative proportion of each product is dependent on, amongst other things, the temperature at which the reaction is carried out.

The preferred conditions for a particular reaction may also be influenced by the form in which the product is needed for the next step. For example it is helpful if product can be isolated from one reaction in the solvent which is needed to provide the reaction medium for the next step.

As the detail of the process is defined, it becomes possible to start chemical engineering design work to provide the appropriate sizes and types of vessels and pipework in which to manufacture, isolate and purify the agrochemical product. This may lead to feedback to process investigation.

As intermediates or by-products that will be formed in the process are identified, they are tested for toxicity to establish any hazards to people operating the manufacturing plant or transporting the substances. Environmental impact studies are also carried out to enable responsible decisions on waste disposal.

PRODUCTION SCALE-UP

Once an acceptable process has been developed on a laboratory scale it has to be scaled-up for large volume production. The scale-up factor may reflect 10 000-fold volumetric increases, involving major changes in other parameters. Reaction times may increase from minutes to hours. There may be 20-fold increases in length of mixing time and 10- or 100-fold reductions in the ratio of surface area between reagents to volume within the reacting mixture.

Processing time can increase for many reasons. Some may simply be related to the physical functions of moving large volumes. For example it takes 1½ to 2 hours to fill a 10-m^3 vessel with water through a 5-cm pipe, compared with a few seconds to fill a small laboratory vessel. The speed of heat movement through a large volume will also be much slower and less even than through a small volume. Much longer processing times than those possible on a laboratory scale can give rise to new problems if, for example, initial or intermediate reagents are unstable.

Slower rates of mixing and diffusion of reagents in large volumes can result in changes in the speed and nature of reactions. These can give rise to a need for changes to other reaction conditions such as reagent concentration. Mixing rates can depend on the flow properties of the reagents, how and where they are added in a vessel, and the agitator systems.

The area of interface between reagents can be an important determinant of reaction rate, in reactions where reagents are contained in two different phases (such as gas and liquid, liquid and solid, or two phases of liquid). This can be affected markedly on scale-up by parameters such as the speed and shape of agitators and the nature of baffles in the reaction vessel.

The final product usually has to be manufactured to a high purity and with clear identification and quantification of impurities to meet the product specifications set by the toxicological testing described in Chapter 5.

The final process only emerges in detail, however, after the scale-up work has been completed. Each problem encountered on the way will require some feedback into new process investigation and progress will be made by a series of iterations.

PRODUCT PURITY

Product scale-up can affect the physical form and purity of the agrochemical active ingredient. Additional purification or clean-up procedures may have to be added to the production process to ensure the required quality of product.

PRODUCT FORMULATION

The purified product which emerges from the manufacturing plant is not suitable for spraying onto crops. In its raw state the active ingredient is unlikely to be in a form which could be dissolved or dispersed in a farmer's spray tank. Even if this were possible, the solution and dispersion would be unlikely to have the ideal biological performance or handling properties.

To convert it into a usable product, the active ingredient has to be formulated into a desirable physical form with appropriate carriers.

The final formulated product should be that which gives:

- optimal biological performance and minimal risk of phytotoxicity;
- optimal safety to users and to non-target species;
- acceptable practical handling properties (e.g., ease of measurement, ease of pouring into a spray tank, ease of mixing in the tank);
- good stability in storage (shelf life) under a range of conditions including sub-zero to tropical temperatures;
- safety in storage, distribution and handling;
- compatibility with containers and packaging;
- compatibility with materials of construction of the application machinery;

— compatibility with other products with which the new product is likely to be mixed in the spray tank (including water which would be of doubtful quality);
— freedom from application problems, such as foam formation in the spray tank or risk of blocking spray nozzles;
— acceptability to registration authorities, including acceptability of all minor, non-pesticidal components;
— suitability for manufacture with available formulation equipment, often in the country of use.

Formulations may take a number of physical forms. The most common types are:

Solids — Water-dispersible or soluble powders
— Soluble or water-dispersible granules
— Granules for application as undiluted solids
— Dusts
— Powders for seed treatment

Liquids — Water-soluble concentrates
— Emulsifiable concentrates
— Oil/water emulsions
— Suspension concentrates
— Micro-encapsulated formulations

The type of formulation chosen for any particular active ingredient will depend upon the chemical and physical properties of that molecule, the type of application method to be used and the relative biological performances of the alternative formulations.

For some situations, liquid products are preferred because they are easy to handle and measure (e.g. in seed treatment machines). Aqueous solutions are only an option for products of high water solubility such as glyphosate or paraquat.

Emulsifiable concentrates are solutions of the active ingredient in a suitable organic solvent, with emulsifying agents to allow the solution to disperse readily in water in a spray tank. Formulations of this type are suitable for substances which are readily soluble in appropriate organic solvents.

Suspension concentrates, or 'cols' are suspensions of very finely ground solid particles in an aqueous medium containing dispersing agents to maintain the stability of the system. These are suitable for solids of very low solubility in water. Careful development is needed to produce a stable suspension which will not settle out to a thick sediment during storage.

Oil/water emulsions contain active ingredient dissolved in an organic solvent which is then emulsified in water (or vice versa) in a high-shear blender.

Suspo-emulsions are mixtures of suspensions and emulsions of water-immiscible liquids.

The simplest solid formulations are wettable or soluble powders and dusts. These contain particles of active ingredient mixed with an inert powder carrier such as clay or calcium bicarbonate. Wettable or soluble powders also contain dispersing agents to encourage dispersion or dissolution in the spray tank.

Soluble or dispersible granules can be made by compressing, extruding or agglomerating powders. They tend to have better handling characteristics and are easier to measure into spray solution than powders or dusts. The solid particles may be micronized before compression in a way that ensures extremely small particle size in spray suspension and consequently better distribution of active ingredient on the final target. Other forms of granules are designed for direct dry application to the treated area in solid form and undiluted.

Micro-encapsulation involves the encapsulation of the active ingredient in minute polymer capsules and dispersing these in water in a similar way to solid particles in a suspension concentrate.

Apart from the active ingredient, formulations can contain dispersing agents, anti-settling agents, anti-foams, anti-freeze, bactericides and inert carriers. They may also include surfactants to aid the wetting of leaf surfaces and the sticking, spreading or uptake of the chemical into plant, fungus or insect. The correct type and concentration of surfactant can play a major role in affecting the biological performance and safety to the crop of an active ingredient. The testing of formulation additives and their concentrations forms an important part of the early biological testing programme.

Wetting agents can markedly affect the area of surface contact between a droplet of spray chemical and the sprayed surface. Electron micrographs of droplets with and without surfactant (Plate 20) illustrate this effect.

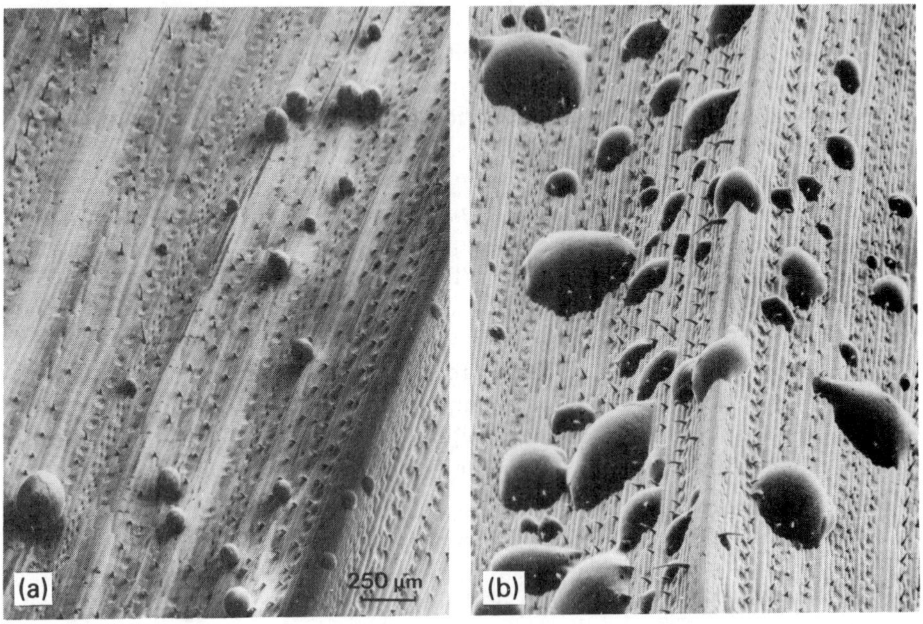

Plate 20 — Scanning electron micrographs showing spray droplets with surfactant (a) and with surfactant (b). The latter show a lower contact angle and spread better on the leaf surface.

Larger droplets can bounce off leaf surfaces if the surface tension of the spray solution is too high. Surfactants can also affect the degree and speed of uptake of the chemical by the plant, fungus or insect by solubilizing the active ingredient on the plant surface and/or by affecting the waxy cuticular barriers to chemical entry.

The physical effects of adding surfactants can, depending on the physical properties of the active ingredients, be reflected in enhanced biological performance.

By altering solubility and penetration into plants, solvents and surfactants can also have a major effect on phytotoxicity. Solvents and surfactants can damage the leaf cuticle and increase the rate of uptake of the chemical spray to a level which is too great for the plant to cope with. Surfactants can also be phytotoxic themselves and a biological testing programme is required to select optimum wetter or solvent rate and type to give the best balance between efficacy and phytotoxicity. Plate 21 is an electron micrograph showing the effects of phytotoxic surfactants on leaf epidermal cells.

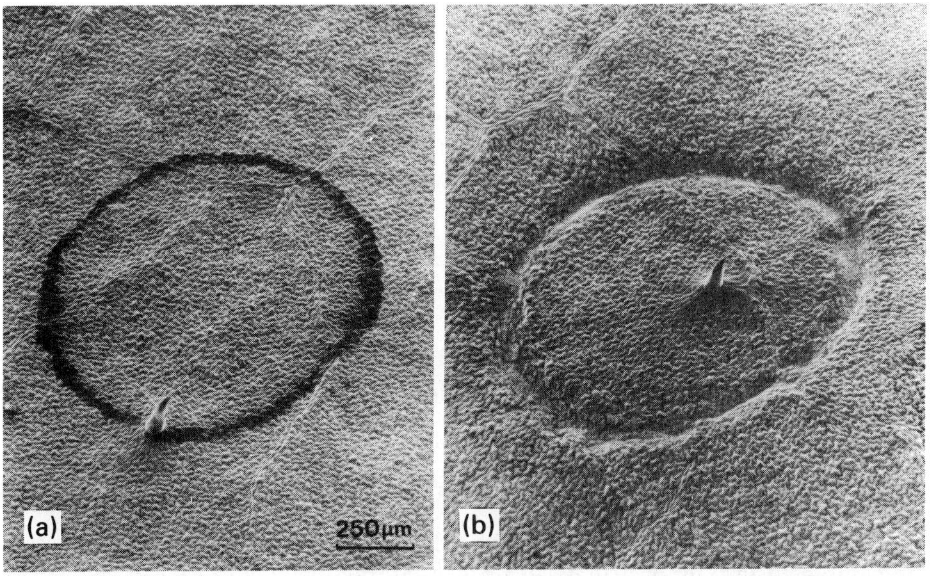

Plate 21 — Scanning electron micrographs showing (a) dried ring deposit from the edge of a surfactant droplet and (b) a ring of collapsed epidermal cells from a phytotoxic spray droplet. The latter would show 'ring-burn' symptoms.

For insoluble active ingredients applied to the plant in particulate form, particle size can often influence biological performance because of the surface area to mass ratio effect. Often, finely milled material presented as, for example, a colloidal suspension dispersed in water with particle size of *c* 2p can be significantly more biologically effective than a large particle size water-dispersible powder product. This is because there are many more particles in a given weight of chemical applied.

These particles will spread more evenly over the plant surface and will dissolve and be taken up more readily.

By correct choice of formulation, useful modifications can be made to product performance. For example, the use of different types of micro-encapsulation can alter the release rate of a chemical in soil or on plant surfaces. Encapsulation polymers can be made to be permeable, allowing slow release of active ingredient, or impermeable, giving delayed release until the capsule ruptures. Capsule wall thicknesses can be controlled in manufacture to vary these effects.

Possible benefits from micro-encapsulation, depending on the active ingredient, are:

- protection from environmental degradation giving, for example, longer periods of pest control;
- reduction in toxicity by providing a protective barrier between the chemical and the operator;
- reduced phytotoxicity;
- enabling co-formulations to be made with otherwise incompatible chemicals.

The choice of formulation will depend to a large extent on the physical properties of the active ingredient and the influence of formulation on biological performance. There are, however, increasing regulatory pressures influencing selection. Many countries, notably within the EEC, are pressing for formulations which do not contain solvents (specifically emulsifiable concentrates) because of concerns about the toxicity of the solvents themselves.

PRODUCT STORAGE TESTING

In parallel with the biological testing of alternative candidate formulations, the products are also subjected to rigorous storage testing to ensure that they will have the necessary shelf life. Because it is necessary to gain a measure of likely long term shelf life quickly, the storage testing procedures include a number of accelerated tests to show the effect of high temperatures (up to 50°C), freezing and cycling temperatures from very cold to hot.

Initially, storage tests are carried out with small quantities, but finally they must be completed in bulk in actual sales packs. Storage testing in sales packs is also necessary to ensure that there is no chemical interaction between the product and its packaging materials.

The final result of the product manufacturing and formulation development process is a product which can be manufactured efficiently and safely, and provided at an affordable price to the farmer. It will be formulated in a way which is easy and safe to handle and gives optimum biological performace. The packaging will be attractive, easy to transport and use, and convenient for disposal.

PACKAGING DEVELOPMENT

The packaging in which crop protection chemicals are distributed and sold have traditionally needed to fulfil a number of objectives:

- To contain the product in a sufficiently robust package to prevent loss through spillage, leakage or permeation during filling, transportation, storage or use.
- To preserve the contents as far as is possible from physical or chemical deterioration.
- To provide a safe and effective means of closure and reclosure.
- To have acceptable appearance and handling characteristics for the customer.
- To have provision for suitable labelling to give information on contents, handling precautions and instructions for use.
- To be economic to produce and fill.

During the product development process a number of test procedures are carried out to ensure that the packs are sufficiently robust to comply with national and international transport regulations. Long-term storage tests are used to check that there are no adverse interactions between the packaging materials and the chemical contents under a wide range of temperature and humidity conditions. Products must be able to withstand extremes of conditions from freezing stores on farms in Canada or Northern Europe in winter, to the heat and humidity of corrugated iron sheds in the tropics.

To meet its needs, the agrochemical industry has pioneered areas of packaging development, for example the development of solvent resistant plastics to replace metal containers. Since the mid-1980s two major areas of concern have come to challenge the industry and will play a crucial part in shaping the future of crop protection packaging. These are the exposure of spray operators and the problem of empty pack disposal.

Whatever the levels of handling hazard associated with the toxicity of the agrochemical product, it is sound practice to minimize the chance of contamination of sprayer operators. One means of reducing exposure is to wear appropriate protective clothing. Packaging can also make a major contribution.

Containers have been redesigned to minimize the risk of splashing during pouring. Soluble packs are being introduced where the chemical is contained within a water-soluble sachet. The sachet is taken from a protective carton and dropped into the spray tank with no exposure of the operator to the chemical inside. The water-soluble film dissolves and is sprayed together with the rest of the spray solution, obviating a pack disposal problem.

Enclosed and refillable bulk supply systems are also being introduced which allow chemicals to be pumped through a sealed line from a bulk container into field sprayers. The bulk container can then be returned to the chemical manufacturer or distributor for refilling.

In some countries disposal of empty containers is becoming increasingly difficult. Packaging materials have been developed which minimize any chemical retention in the container. For example polyethylene terephthalate (PET) has been developed as a plastic for bottle manufacture which has minimum chemical retention properties. Packs have also been designed with no nooks and crannies to retain small quantities of product.

Some waste disposal sites, however, will not accept pesticide packs even if they have been thoroughly rinsed. These pressures favour packaging which can be easily destroyed on the farm, such as fibreboard cartons, or can be recycled. The preferred choice of packaging will play a major part in the choice of formulation development, and packaging strategy must be an integral part of the development plan for a compound.

It is clear that increasing attention will be paid to packaging issues by chemical producers and by legislative authorities and that this will be an area of significant change.

Plate 22 — Enclosed system for transferring chemical from container to spray tank.

Plate 23 — Refillable bulk container for chemicals.

7

Application equipment

THE EARLY DAYS

The earliest uses of agrochemicals in Europe, seed soaking treatments with copper sulphate and lime for bunt control on wheat, required no more than a large steeping tank for their application. Methods of applying pest control chemicals to field crops were first needed in the middle of the nineteenth century to apply sulphur dust to vineyards for powdery mildew control and liquid Bordeaux mixture for downy mildew. The earlier form of dusting machine consisted of a pair of bellows to blow clouds of sulphur dust onto the plants. Millardet's first suggestion for applying Bordeaux mixture was to walk between the vine rows with a bucket in one hand and switch of leather in the other, flicking mixture to right and left!

The first pressure sprayers to be used to apply liquid agrochemicals were garden syringes which drew chemical solution from buckets. These were clearly impracticable for spraying large areas of crops. Strong incentives to develop better techniques and equipment came from French vineyard owners who were trying to combat downy mildew. Invention was also actively encouraged by the French regional agricultural societies, who organized competitions with prizes for the best and most easily used machines (Large, 1940).

One of the first field-scale sprayers, invented by Armand Cazenove of La Réole, had a handle-driven revolving brush operated through a train of gear-wheels (Fig. 24). The brush was fed with spray mixture from a portable copper tank to which it was attached. As the handle was turned a stirrer went round in the tank, the brush revolved in its trough and a scraper pressed against the bristles flicking off a fine spray.

Other implement makers, village blacksmiths and brassworkers, invented other types of sprayer most of which were essentially prototypes of the modern knapsack and hydraulic pressure sprayers. They consisted of a brass or copper tank, a hand operated pump, a length of rubber tube running from the tank and a spray lance fitted with a spray nozzle (Plate 24). Some of the best early models came from a brassworker called Vermorel of Villefranche whose products soon became known worldwide. His highly successful knapsack sprayer had the advantages of excellent

Fig. 24 — Cazenove's revolving brush sprayer.

nozzle design and a diaphragm pump which meant that no moving piston parts were in contact with the Bordeaux mixture. There was very little to go wrong. Vermorels' sprayer was developed further in the USA where, by the late 1880s, C. V. Riley had produced a sprayer, called the 'Eureka', using a new cyclone jet.

As the use of Bordeaux mixture expanded from vines onto potatoes (for potato blight control) in the 1880s, larger hydraulic pressure sprayers were produced which could be pulled through the fields by horses (Plate 25). Horse-drawn sprayers were also developed in the USA for use in orchards. It is estimated that between 1887 and 1892 the number of fruit tree spraying machines in the USA rose from about 50 to 50 000.

Since those early days, application technology has evolved and improved. Nevertheless, the majority of spray equipment still follows the same general principles.

APPLICATION METHODS

Agrochemicals can be applied to their targets in a range of physical forms. These are:

— sprays where the active ingredient is dissolved or dispersed in oil or water,
— dusts,
— dry granular products for application undiluted,
— baits for pests such as rats and mice,
— fogs or mists for spore treatment.

Plate 24 — Early brass knapsack sprayer.

The choice of application method will depend on the nature and location of the target. It will also be affected by the physical mobility of the agrochemical active ingredient in the plant, and its mode of action on the pest.

The formulation of the chemical will be designed to fit with the application method. Typical types of target to be covered with agrochemicals are:

— crop stems, fruit and foliage for insect or fungal disease control or crop growth regulation;
— weed leaves and stems for weed control;
— soil for pre-emergence weed control, or for systemic uptake of fungicides or insecticides;
— soil, for the control of pests living in the soil;
— seeds before planting;
— transplant seedlings before transplanting;

Plate 25 — Horse-drawn sprayer.

— flying insect bands (for example locusts);
— selective control of pests which can be attracted to point sources of treated bait.

The appropriate application equipment depends upon the physical form of the material to be applied and the shape, size and nature of the target. The location of fungal or insect pests in the crop's canopy also affects the application techniques required, to ensure the correct contact between chemical and target.

APPLICATION OF SPRAYS TO LEAVES, FRUIT OR STEMS

To control many weeds or pests and to achieve some plant growth regulation effects, the chemical active ingredient must be deposited in an even film over the surface of the treated plant.

The extent to which all surfaces need to be evenly covered depends upon the nature of the pest and the mobility of the chemical on and in the plant. For example, protectant fungicides such as the dithiocarbamate or phthalamide compounds protect plants from germinating fungal spores by forming a fungitoxic chemical layer over leaf surfaces. These compounds have little or no mobility on or in plants and, therefore, the effectiveness of the fungi-toxic film is largely dependent upon the uniformity of its application. Unprotected areas of leaf surface would allow fungal infection to penetrate and take hold. In practice all chemicals show some degree of redistribution on leaf surfaces or in leaves to compensate for 'gaps'.

Chemicals which are used to control pests on the lower surfaces of leaves but which do not move readily through the leaf tissue from the upper to lower surface (for example the pyrethroid insecticides) must be applied in a way which will enable good coverage of the underside of leaf surfaces. By contrast, chemicals which have excellent mobility within plants may not be particularly sensitive to application method. The phloem-mobile herbicide glyphosate, for example, can be applied to part of a leaf surface and will enter that leaf and be translocated down the plant, achieving effective kill of the whole plant. Many fungicides and insecticides are also systemic and are able to redistribute within treated plants. For these, uptake into the plant is critical rather than surface distribution; this is an issue considered in the section on formulation in relation to the use of additives to improve surface wetting and penetration.

In general the development of spray equipment has sought to improve the efficiency of spray coverage into the crop canopy while, at the same time, making applications easier for the operator and minimizing the risk of spray drift into non-target areas.

Efficiency of spray coverage

The efficiency of spray coverage depends on

— volume of spray,
— droplet size,
— droplet velocity in relation to the target,
— the angle of the target surface in relation to the spray cloud (fluttering leaves will generally receive better coverage than still areas),
— the density of crop canopy and the degree of shelter afforded by one leaf for another.

Each of these factors is considered in more detail.

Volume of spray solution

The rate of recommended use for an agrochemical is usually defined in terms of grams of active ingredient per hectare (g ai/ha). The chosen rate of active ingredient may be applied in a range of volumes of diluent (usually water). Typical spray rates for diluted solutions are classified into high, medium, low, very low or ultra-low volume (usually abbreviated to HV, MV, LV, VLV and ULV). Rough orders of magnitude for each of these are given in Table 13.

Table 13 — Typical spray volumes (l/ha)

	Field crops	Trees and bushes
High volume	>600	>1000
Medium volume	200–600	500–1000
Low volume	50–200	200–500
Very low volume	5–50	50–200
Ultra-low volume	<5	<50

In recent years the trend has been towards decreasing volumes of spray per hectare to ease the problems and inconvenience of carting large volumes of water, and to reduce the time needed for farmers to make the spray applications. The actual volume chosen for a particular spray application will depend also on the size of the target (particularly for tree and bush crops). Volumes may in fact be changed during the course of a season to reflect changes in target area. It is conventional practice in French vineyards to raise the volumes of application from 200 or 300 l/ha at the beginning of the season, when vines are small following pruning, to 1000 l/ha for the later sprays when the crop foliage canopy is fully developed, sprawling and dense.

When pesticides are used on crops which can have a wide range of sizes of foliage canopy per hectare, e.g. fruit trees, recommended rates of active ingredient application need to reflect that canopy size as well as the area of ground covered. In these circumstances recommendations are often made for chemical to be applied in certain concentrations (e.g. g ai/hl of spray solution) assuming 'high volume' practice and allowing rates of chemical per hectare to vary with application volume. If growers choose to use low or medium volume application, then concentrations are adjusted accordingly.

Droplet size

The optimum volume of spray solution is also affected by the droplet size distribution in the spray. The smaller the size of each droplet, the larger the surface area of spray deposit per unit volume and the better the coverage of the target, i.e. the higher the number of droplets per unit area of target covered at a particular spray volume. Smaller droplets, however, are more prone to drift. Where control of drift is critical, larger droplets should be used. Spray equipment which minimizes volume relies to a large extent on being able to produce small and relatively uniform droplet sizes. This is discussed later with reference to controlled droplet application (CDA). The most common measure of droplet size is volume median diameter (VMD), measured in micrometres (microns, μm). A broad classification of sprays according to droplet size is given below.

VMD of droplets (μm)	Droplet size classified
<50	Aerosol
51–100	Mist
101–200	Fine spray
201–400	Medium spray
>400	Coarse spray

Equipment for applying spray

Conventional pressure sprayers for applying liquids are manufactured in a wide range of models to suit the scale of application and the nature of the target. Sprayers range from small hand-operated sprayers for spot treatments (Plate 26) to aircraft-mounted sprayers for very large area applications (Plate 27).

Plate 26 — Knapsack sprayer being used for weed control in smallholder maize.

Plate 27 — Aerial spraying.

Ground-based sprayers may be tractor-drawn or self-propelled (Plates 28 and 29). Different designs are used for different targets. Long boom sprayers are used for spraying arable crops such as cereals or sugar beet (Plate 30). Boom sprayers shaped to fit over vine trellises may be used for applying fungicides or insecticides to grapes (Plate 31). Air-blast sprayers can be used to treat the canopies of trees (Plate 32).

Plate 28 — Tractor-drawn sprayer.

The operating principles of the commonest forms of small sprayers have altered little since the late nineteenth century although the construction materials have changed, generally from brass to plastic. The spray liquid is pressurized by a piston or diaphragm pump, and pumped from the carrier tank through a flow control device and a length of hose to spray nozzles where it is atomized into droplets and propelled under pressure to its target. Some smaller sprayers (such as those common for garden use) use plunger pumps to pressurize the spray tank. Tractor-mounted or self-propelled hydraulic sprayers follow a similar principle, with liquid pumped by a power pump through a pressure control device to the nozzles mounted on the boom.

Air-carrier sprayers, where spray droplets are propelled to their targets in large volumes of high velocity air, are particularly useful for getting spray coverage into the canopy of trees. The droplets may be preformed by hydraulic or centrifugal nozzles before entering the airstream or may be formed by the airstream itself. A typical air-blast sprayer is illustrated by Plate 32.

Plate 29 — Self-propelled row crop sprayer.

Plate 30 — Long spray boom for arable crops.

Plate 31 — Vineyard sprayer.

Plate 32 — Air-blast orchard sprayer.

Small air-carrier sprayers, referred to as motorized knapsack mistblowers, are often used for space spraying, particularly against public health pests, where clouds of small droplets are required for good control.

Controlled droplet application (CDA)
Some machinery development has aimed to reduce spray volumes and optimize the size of spray droplets. The most common form of controlled droplet applicator uses a rapidly spinning serated disc to form droplets through centrifugal forces at the discs perimeter (Plate 33). The movement of the spray droplets formed from the disc may be enhanced by a fan to propel them forcibly to their target.

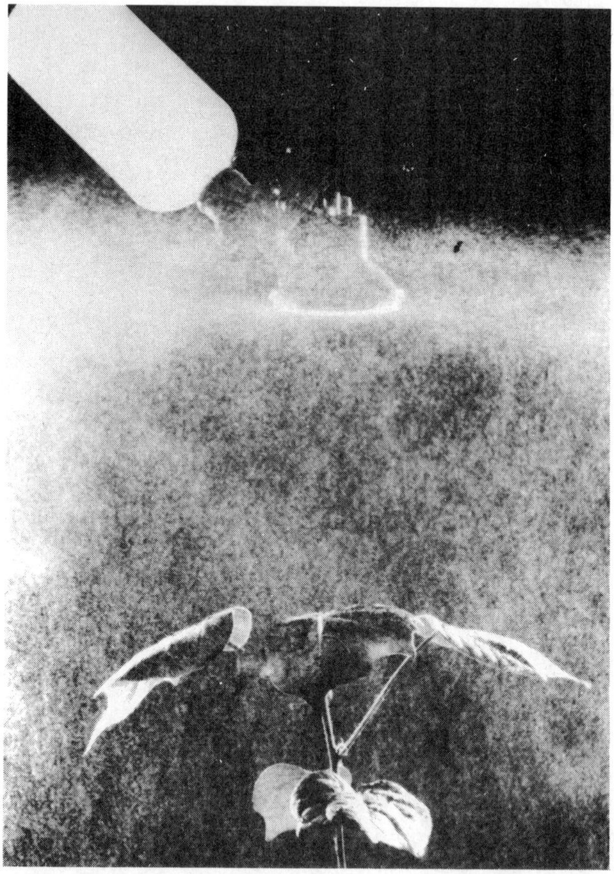

Plate 33 — Spinning disc sprayer producing fine, uniform-size droplets.

The latest advance in spray technology is the use of electrostatics to produce charged droplets of controlled size. The advantage of charged droplets over uncharged sprays lies primarily in their ability to wrap around the target, for example covering the underside of foliage. An electrostatic 'Electrodyn' sprayer and the deposition patterns produced are illustrated in Plates 34 and 35. Plate 34 shows

Plate 34 — Charged spray droplets from an 'electrodyn' sprayer.

droplets following electrostatic fields of force between the charged spray nozzle and the plant foliage. The smaller the droplet sizes and the lower the volumes of spray, the less suitable does water become as the carrier because of its rate of evaporation, particularly in warm climates. An oil-based spray is preferable for CDA in warm climates. For electrostatic spraying, oil or solvent carriers are needed to provide the correct electrical properties.

INJECTION

For some agrochemical uses it is necessary to place applications accurately. Injection equipment has been developed for putting chemicals into soil and into the trunks of

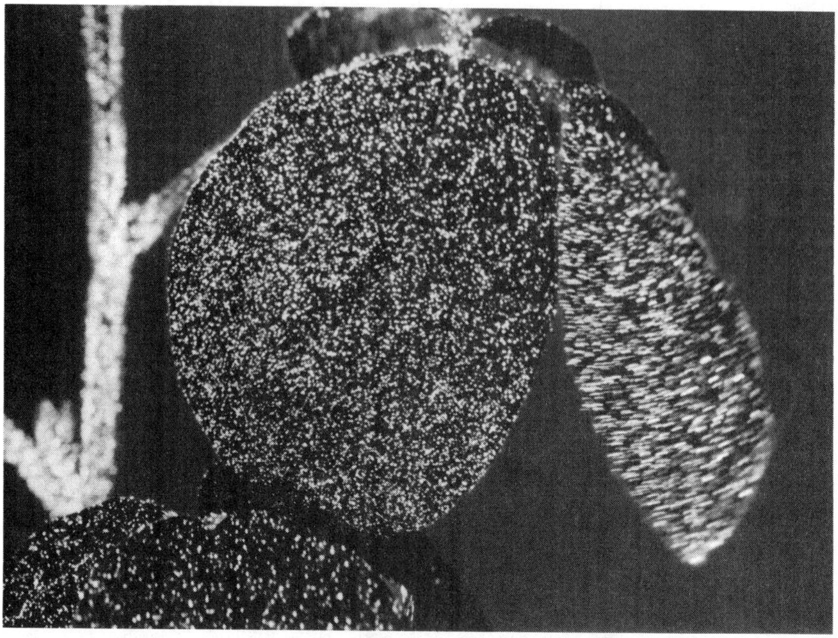

Plate 35 — Droplet deposition pattern from an electrostatic sprayer showing coverage of the undersides of leaves and of stems.

trees. Soil injection equipment is used to apply volatile soil fumigants for the control of nematodes beneath the soil surface. The principles of the equipment are similar to those for spray application, except that the chemical is led from the metering device down to outlet jets below the soil surface. These may be contained in the back of a cultivator tine, where a band of chemical is to be applied below the surface of the soil (Plate 36) or in a single soil probe for isolated point applicators (Plate 37).

Injection equipment has also been developed for forcing chemicals into the trunks of trees. Two practical uses have been for the control of tree diseases such as Dutch elm disease (caused by the fungus *Ceratocystis mulmi*), and the application of tree growth control chemicals. Holes are drilled into the ring of xylem vessels which lie beneath the bark and probes are inserted from a pressure pump. Chemical pumped into the tree is then carried in the upward flow of water through the xylem vessels and distributed through the tree.

DUSTS

Dusts are becoming progressively less important as an agrochemical formulation because of their bulk when formulated, and the greater problems of drift and of inhalation by sprayer operators. For most uses they have been replaced by sprays. The application of dust seed treatment, is considered a special topic. Admixture with stored grain for the control of insect pests is achieved by very simple means such as dusting bags, or mixing with a shovel!

Plate 36 — Soil injection equipment attached to a cultivator tine.

DRY GRANULES

Dry granular formulations of agrochemicals are used in a number of very different situations. In developed agriculture, granules are often the preferred formulation for control of soil insect pests. They are applied in bands along the line of planting or in a furrow with the seeds. Plate 39 shows a granule applicator used for placing granules in furrows with maize seeds for corn rootworm control.

Granules can also be broadcast onto the surface of the soil, common practice for control of pests in turf grass, or into the flood irrigation water in rice paddy field. Granule applicators usually consist of a hopper box, into which the granules are poured, a metering device to ensure the application per hectare, and a delivery system to place the granules where needed. Granular applications, because they must achieve an adequate number of point sources of agrochemical per hectare without use of further diluents, tend to require relatively large quantities of formulated product per hectare, often 30–45 kg product/ha compared with 100 g–1 kg for a dilutable product. The cost of granular application is, therefore, often higher than that of sprays. Requirements of a good granule applicator are that it should

— deliver accurate amounts of product per hectare as calibrated,
— spread particles evenly,
— avoid damage by grinding or impaction,
— give good feeding of material to and from the metering device,
— be robust and proof against moisture and abrasion.

Plate 37 — Soil injection through a pressure probe.

SEED TREATMENTS

Seeds are often treated with fungicides and insecticides to protect them from pest attack through the germination, seedling and early establishment phases of its growth. Formulations used may be liquids or dry powders. The seed treatment machine needs to:

— ensure accurate metering of formulation quantity per unit of seed,
— ensure a good and uniform coverage of the seed with the agrochemical,
— allow a sufficiently rapid throughput of seed.

A number of different mixing principles have been adopted in commercial equipment. Examples are augers and centrifugal drums (Plate 40).

Plate 38 — Injecting chemical into a tree trunk.

FOGS

Fogs or mists are produced when aerosol droplets of less than 15 mm diameter are suspended in air. Pesticide fogs are used for insecticides or fungicides when the objective is to fumigate an enclosed space, controlling pests in inaccessible cracks and crevices or flying in the air in the enclosed space. Common sites for fogging are glasshouses, warehouses, ships' holds and farm sheds or storage silos. For this technique, formulations of chemicals are usually dissolved in an oil of suitably high flash point. The fog is produced by vaporizing oil from a hot surface on the machine, or metering it into a hot exhaust gas stream from an engine, and expelling it into the atmosphere where it condenses again to form very fine suspended droplets. Equipment is illustrated in Plate 41.

(a)

(b)

Plate 39 — Equipment for applying granules in furrows (a) Side view. (b) Rear view showing a tube for delivering grannules into a band above planted seeds.

(a)

(b)

(c)

Plate 40 — 'Rotostat' seed treatment machine. (a) In use in a seed treatment mill. (b) External view. (c) Cut-away view showing interior of centrifugal treatment mechanism

Plate 41 — Fogging machine in use for controlling public health pests.

8

Financial analysis and product planning

AGROCHEMICALS AS AN INVESTMENT

As the preceding chapters have shown, the agrochemicals industry is dependent for its survival and growth on successful commercial exploitation of research and development. Research and development work is needed to invent new products and bring them to the market, to improve the effectiveness and profitability of existing products and to satisfy the ever-growing human and environmental safety requirements which society and regulators demand of new and existing commercial products. As Chapter 9 will show, however, this exploitation has to take place against a background of increasing industrial maturity. More and more good products are available and the majority of major agrochemical needs are served to a greater or lesser degree of perfection by products which now exist. The boards of companies within the industry are therefore forced, under increasing pressure, to consider very carefully both how much to invest in research and which projects should be given priority.

For the industry as a whole, the turnover or gross income from sales is apportioned roughly as shown in Fig. 25. About 40% of the ex-manufacturer sale price of a product is represented by the variable costs of manufacture, formulation, packing and distribution; a further 20% is spent on costs of sales; promotion, sales staff, and sales administration etc. This leaves 40% to pay fiscal expenses (about 10%) reward fixed capital investment, to fund further research and development and to return a cash surplus.

At present the industry is R & D intensive, investing about 8–10% of its sales revenue on R & D. In comparison with other industries this ranks near the top, slightly behind pharmaceuticals (*c* 11%), electronic data processing (*c* 13%), and aerospace equipment (*c* 16%). This contrasts with an average R & D spent for all manufacturing industry of *c* 2% of sales (Table 14).

Of that research expenditure, there is probably an approximately equal allocation between improving or sustaining existing products, developing new, identified products, and inventive research. Of the inventive research expenditure, about 70–80% will tend to be in 'known science' leading to conventional product areas, and 20–30% in higher risk 'novel science' in an attempt to create new types of business.

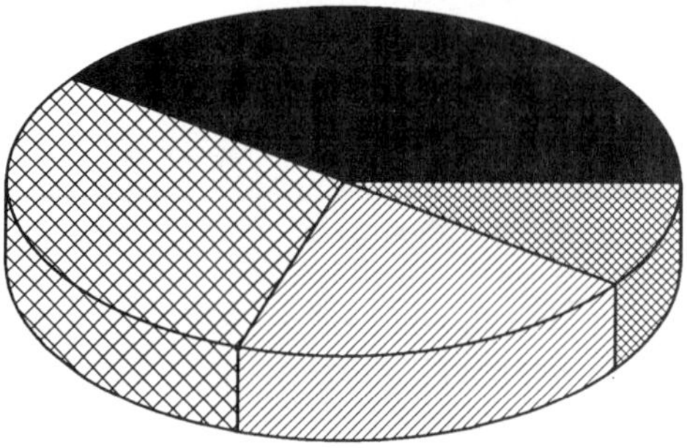

Fig. 25 — Approximate apportionment of corporate income for a typical agrochemical company.

Table 14 — Proportion of sales income spent on R & D (examples based on Business Monitor MO14 1981 HMSO)

	Intramural expenditure on R & D as percentage of sales
Electronic capital goods	26
Aerospace equipment and repair	16
Electronic data processing	13
Pharmaceuticals	11
Agrochemicals†	9
All chemical industry	3
All manufacturing industry	2
Mechanical engineering	1
Iron and steel	0.4

†Agrochemical industry sources.

ISSUES DETERMINING LEVELS OF INVESTMENT IN R & D

It is not possible to define a correct level of agrochemical R & D investment. Levels of investment are set in practice as a balance between those which each company believes that it can afford and those which could be justified by the potential rewards.

There is also a reflection of the spending patterns of other major competitors with a measure of 'herd instinct'. The level which can be afforded is often much clearer to see than is the level which can be justified.

The level which can be afforded is set by the projcted profitability of the forecast existing business, and the cash which that generates, coupled with the borrowings which borrowers and lenders deem prudent. At the current state of maturity of the agrochemicals industry (see Chapter 9), most companies will want to finance R & D from existing income and not with borrowings. Very new and potentially exciting businesses in new technology areas can escape this constraint only if they can convince suppliers of speculative venture capital, or if they can be sustained under the wing of a large and profitable parent business. Biotechnology R & D is currently financed by one or the other of these means in the expectation of a large 'break-through' (see Chapter 10).

As industries mature, they tend to reduce in profitability and are able to finance proportionally lower R & D expenditure. Within the total budget, allocations of resource have to be made between categories of research: support of commercial products, new product developments or 'inventive' research. The relationship between research and economic reward is clearest for work on existing commercial products. For example, programmes of work to maintain existing registrations or to reduce production costs for established products can have clearly defined and often highly predictable outcomes whose financial value can be easily calculated. Usually research projects of this type also have a rapid financial payback.

In the current regulatory climate, discussed in Chapter 5, government authorities are increasingly demanding new information to bring product safety dossiers up to the latest standards, or to satisfy new areas of concern. If this information is not provided within an agreed (and usually short) period of time, the registration could be cancelled. The financial benefit of re-registration can be assessed for well-established and relatively predictable business on the basis of gross margin and profit contribution. For many products which are well up, or near the top of their sales growth curve, such protective R & D expenditure is rewarded within one or two years. The rate of return on investment is very high, relative to other directions of research expenditure, making it a high priority for most agrochemical companies. It is, of course, even more of a financial priority because loss of important registrations could well result in reductions in short and medium term cash flow and gross profit from which to fund longer term R & D.

Research to improve the profitability of existing products can also be very attractive. Profitability can be improved by reducing production costs, enhancing products to enable them to move into higher priced market sectors or expanding their application into new areas of business. For research which increases profit margin, the added income is immediately geared across the total value of product sold. An increase of 10% margin on a product with £50m turnover is immediately worth £5m per annum. Risks are lower than for new product development because the areas of investigation have a much more predictable outcome.

Research to develop new products which have been sufficiently well characterized to enter the development phase of R & D, as explained in Chapter 4, is riskier than established product work but is still amenable to a reasonably accurate financial appraisal.

The risks for new agrochemical product development are both technical and commercial. Technical risks exist in all areas of investigation. In biology, it is not known until an advanced stage of development whether the candidate compound will reliably perform all that its use specifications assumed; that is to say controlling the target range of pests at the required level of efficacy at the predicted rate of application without adverse effects on the crop. In toxicology and environmental studies, the safety of the compound is not known with confidence until the end of the testing programme. Until manufacturing processes are completely defined, production cost is still subject to uncertainty.

Commercially, there are risks associated with uncertainty of price and performance as compared with the competition and other areas of ignorance. This is compounded by the lack of detailed characterization of the candidate product at an early stage, and the inability to predict in detail the commercial environment of the marketplaces around the world that the product will enter and (it is hoped) grow into in seven to fifteen years time. Nevertheless, once one is able to define the broad commercial opportunities open to a conventional pesticide product, it is possible to make estimates of:

— the rough area in hectares which will be treated in each target market at maximum sales; estimates can also be made of the number of hectares treated annually as the product's use grows to its maximum and subsequently declines;
— the likely rate of chemical application;
— the likely price of application per hectare that the market will stand, and from this the price of the product itself;
— the likely cost of production;
— costs of marketing and sale;
— costs of development.

These factors can then be combined into a discounted cash flow analysis showing the financial profile of the project. An example of a typical cash flow analysis is given in Table 15. This shows the annual sales, gross margins, selling costs, development costs, working capital employed and annual profit from the project. It also shows the rate of return which the project earns on investment over its life and the project's present value. Few of these parameters can be predicted with a high level of certainty. Analysis can, however, explore a range of assumptions for each parameter to demonstrate to management where the greatest financial risks lie, and to help make judgements about the wisdom of proceeding with the project; taking into account also the range of the technical risks. (This book is not intended to be an introduction to project analysis or accounting, and so readers who want to gain a better understanding of cash flow analysis should consult more specialist sources.)

Financial analysis is very much more difficult for projects in novel areas of research. A policty statement that *x*% of R & D funds will be spent on 'novel' research represents a commitment to the need to refresh the business in the longer term with products which cannot yet be clearly conceptualized. it does not, however, give any guidance as to whether the funds are likely to be a wise investment or a 'shot in the dark'. Such investment must be regarded as 'casting bread upon the waters' (Ecclesiastes II,1).

Table 15

		£ million of year													
		1989	1990	1991	1992	1993	1994	1995	1996	1997	1998	1999	2000	2001	2002
Volume (TE AI)		—	—	.6	4.2	9.1	16.3	33.0	49.5	61.2	71.0	80.2	88.1	88.9	90.2
Sales (Giv)		—	—	.1	1.0	2.6	4.2	7.8	12.2	16.3	18.6	22.5	25.8	26.2	26.7
Distribution costs		—	—	—	.1	.2	.4	1.1	1.9	2.5	3.0	3.6	4.2	4.2	4.3
Active ingredient cost		—	—	—	.1	.1	.2	.5	.7	1.0	1.2	1.4	1.7	1.8	1.7
Formulation and packing		—	—	—	—	—	.1	.3	.5	.7	.9	1.1	1.3	1.3	1.4
Group gross margin		—	—	.1	.8	2.2	3.4	6.0	9.1	12.1	13.6	16.3	18.7	18.9	19.2
Marketing		—	—	.01	.06	.10	.19	.33	.77	1.31	1.65	2.10	2.59	2.63	2.66
Launch		—	—	.04	.10	.23	.17	.38	.73	.74	.07	.07	.08	.08	.09
Development		.02	1.41	1.47	.84	.69	.13	.10	.07	.02	.02	.01	.01	—	—
Country contribution		−.02	−.1.41	−1.39	−.20	1.17	2.96	5.16	7.50	10.01	11.83	14.14	16.01	16.15	16.45
Fixed production expense		—	—	—	—	—	—	.01	.01	.01	.01	.01	.01	.01	.01
Central R&D		—	.69	.50	.53	.25	—	—	—	—	—	—	—	—	—
Contribution to group profit		−.02	−2.10	−1.89	−.72	.92	2.96	5.15	7.49	10.00	11.82	14.13	16.00	16.14	16.44
Working capital															
Total group stock		—	—	.02	.06	.13	.26	.58	.91	1.24	1.54	1.86	2.13	2.17	2.24
External debtors		—	—	0.2	.25	.80	1.22	1.96	2.99	3.96	4.55	5.38	6.20	6.27	6.36
Creditors		—	−.20	−.20	−.18	.19	−.16	−.33	−.52	−.67	−.71	−.87	−1.01	−1.04	−1.06
Total working capital		-	−.20	−.16	.13	.74	1.31	2.20	3.39	4.53	5.38	6.37	7.33	7.40	7.54
Cash flow															
Contribution to group profit		−.02	−2.10	−1.89	−.72	.92	2.96	5.15	7.49	10.00	11.82	14.13	16.00	16.14	16.44
Movement in working capital		—	.20	−.04	−.29	−.61	−.57	−.89	−1.19	−1.14	−.85	−1.00	−.95	−.07	−.14
Fixed capital — new		—	−.50	—	—	—	—	—	—	—	—	—	—	—	—
Cash flow (before tax)		−.02	−2.41	−1.93	−1.01	.31	2.39	4.26	6.31	8.86	10.97	13.14	15.04	16.07	16.29
Taxation		.01	.60	.83	.41	.06	−.73	−1.36	−2.09	−3.03	−4.08	−4.59	−5.37	−5.80	−5.84
Cash flow (after tax)		−.01	−1.81	−1.11	−.60	.37	1.65	2.89	4.22	5.83	6.90	8.54	9.67	10.27	10.45
Cumulative cash flow		−.01	−1.69	−2.61	−3.44	−3.13	−1.59	1.13	5.13	10.68	17.28	25.52	34.87	44.82	54.95
Summary statistics															
Group gross margin %		—	—	87	83	85	83	76	75	74	73	73	72	72	72
Group contribution %		—	—	—	—	36	71	66	62	61	63	63	62	62	62
NPV 20% (after tax)	8.19														
NPV 15% (after tax)	13.34														

NPV=Net present value.

For targeted 'applied' research, which dominates the 'novel' end of corporate R & D as opposed to 'fundamental' research, some financial logic can be used which can be helpful in decision-making despite the difficulty of doing it. Generally, organizations will be unwilling to spend too high a proportion of the R & D budget on more speculative areas because the risks are higher and any payback is inevitably very much longer term than that for 'near market' R & D. However, they must spend enough to ensure a long term future. Some guidance on the decision can be gained by considering the cash flow analysis in reverse, and making judgements about how large and profitable a discovery needs to be in order to justify a certain element of R & D expenditure. One can then judge the probability of a discovery of the required value being made in the assumed time scale.

For example, if it were decreed that investment in R & D over a 20-year period should yield a certain level of internal rate of return, assumptions about annual research expenditure, product discovery time and product development, launch production and marketing expenses, taken together with assumed product growth curves can lead to estimates of the size of product discovery needed. This can be illustrated by a worked example (Fig. 26). Here, *A* is the assumed element of

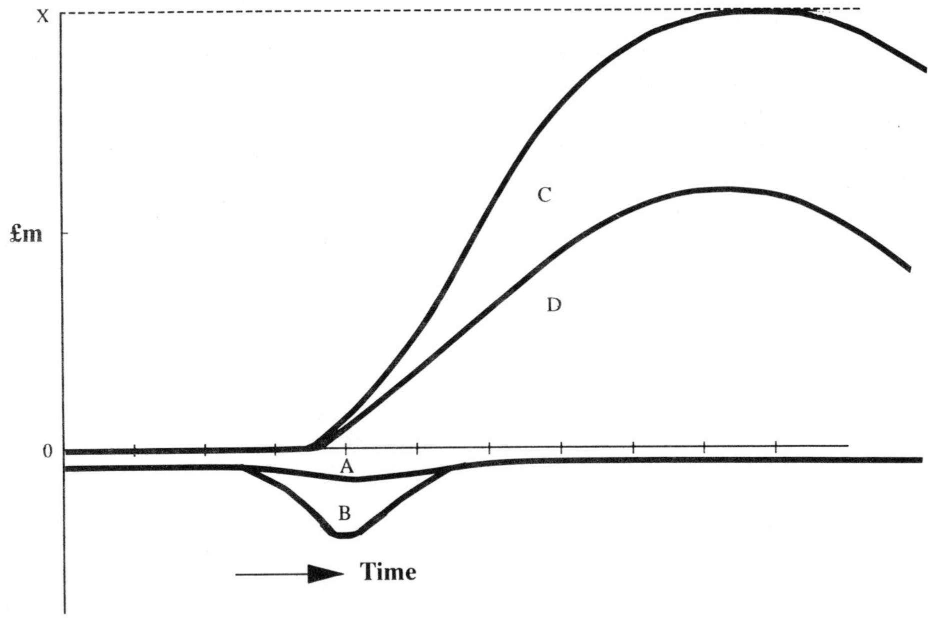

Fig. 26 — Illustrative cash flow model.

research and development costs to discover, develop and support the novel 'target product', *B* is the assumed capital investment to manufacture the 'target product'; *C* is the production and selling costs for the 'target product'; *D* is the returns to reward investment in A, B and C. The present value of D must be equal to or greater than the present value of $A+B+C$. It is possible, therefore, on the basis of any particular set of assumptions, to calculate the minimum total sales value needed (£x) by year y.

If it is assumed that percentage gross margin and percentage direct costs are uniform throughout the life of a product, and that working capital constitutes a constant proportion of sales value, it is straightforward to calculate the present value of contributions as a function of net realization at maximum sales.

For example, Table 16 gives two sales profiles:

(a) a slower growth example growing linearly to maximum sales in 10 years, and subsequently declining,
(b) a faster growth example peaking between years 6 and 8 of sales.

For the slower growth profile, the present value (at 10% discount rate) of the stream of net realizations is $3.97x$, where x is the annual realization at maximum sales. The present value of gross margin is $3.97xy$ where y is the percentage gross margin. The present value of the working capital flows is $0.13x$ and the present value of selling costs is $3.97xz$, where z is the percentage direct selling costs.

Taking the Group average gross margin of 55% and Group direct costs of 19% of sales, the present value of contributions is $1.3x$. By the same procedure the present value of contributions from the faster growth profile is $1.44x$.

Taking the faster growth rate example, the present value of contributions of $1.44x$ is the value as seen from the year before first sales. The value of the contributions as seen from the time of discovery of a new compound will reflect a further period of discounting. For an optimistic six years to first sales, the $1.44x$ will be discounted by 0.62, giving $0.89x$.

If the present value of contributions is to equal the present value of costs of development, estimated for the larger development products to be £30m, annual sales at maximum x, must be at least that given by:

$$0.89x = £30\text{m}$$

or

$$x = £34\text{m}$$

A further allowance must then be made for that contribution required to reward fixed capital investment.

Greater optimism about percentage gross margin would reduce the required realization; for example, the rapid build-up profile and a 60% gross margin lead to a requirement for an annual sales realization of £29m at maximum to repay £30m PV of cost.

The assumptions to be used in the base model can be drawn from historical analogies. Usually there is a range within which values for each parameter are likely to fall and sensitivity analyses can be conducted to demonstrate the effect on the decision of alternative values within that range.

The results of this form of modelling will not define accurately what the financial

Table 16 — Growth assumptions for net sales realization and working capital for the worked example

Year of Sales	1	2	3	4	5	6	7	8	9	10	11	12	13	14	15
Slower growth example															
Sales realization	0.1*x*	0.2*x*	0.3*x*	0.4*x*	0.5*x*	0.6*x*	0.7*x*	0.8*x*	0.9*x*	1.0*x*	1.0*x*	0.9*x*	0.8*x*	0.7*x*	
PV at 10%	0.09*x*	0.17*x*	0.23*x*	0.27*x*	0.31*x*	0.34*x*	0.36*x*	0.38*x*	0.38*x*	0.39*x*	0.35*x*	0.29*x*	0.23*x*	0.18*x*	=3.97*x*
Working capital @ 34%	0.034*x*	0.068*x*	0.102*x*	0.136*x*	0.170*x*	0.204*x*	0.238*x*	0.272*x*	0.306*x*	0.304*x*	0.304*x*	0.306*x*	0.272*x*	0.238*x*	0
Incremental wkg. cap.	0.034*x*	0.034*x*	0.034*x*	0.034*x*	0.034*x*	0.034*x*	0.034*x*	0.034*x*	0.034*x*	0.034*x*	0	−0.0.34*x*	−0.034*x*	−0.034*x*	−0.238*x*
PV of incremental working capital	0.031*x*	0.028*x*	0.026*x*	0.023*x*	0.021*x*	0.019*x*	0.017*x*	0.016*x*	0.014*x*	0.013*x*	0	−0.011*x*	−0.010*x*	−0.009*x*	−0.057*x* =0.13*x*
Rapid growth example															
Sales realization	0.1*x*	0.2*x*	0.4*x*	0.6*x*	0.8*x*	1.0*x*	1.0*x*	1.0*x*	0.9*x*	0.8*x*	0.7*x*	0.6*x*	0.5*x*	0.4*x*	
PV at 10%	0.09*x*	0.17*x*	0.30*x*	0.41*x*	0.50*x*	0.56*x*	0.51*x*	0.47*x*	0.38*x*	0.31*x*	0.25*x*	0.19*x*	0.14*x*	0.10*x*	=4.38*x*
Working capital @ 34%	0.034*x*	0.068*x*	0.136*x*	0.204*x*	0.272*x*	0.34*x*	0.34*x*	0.34*x*	0.306*x*	0.272*x*	0.238*x*	0.204*x*	0.170*x*	0.136*x*	0
Incremental wkg. cap.	0.034*x*	0.034*x*	0.068*x*	0.068*x*	0.068*x*	0.068*x*	0	0	−0.034*x*	−0.034*x*	−0.034*x*	−0.034*x*	−0.034*x*	−0.034*x*	−0.136*x*
PV of incremental working capital	0.031*x*	0.028*x*	0.051*x*	0.046*x*	0.042*x*	0.038*x*	0	0	−0.014*x*	−0.013*x*	−0.012*x*	−0.011*x*	−0.010*x*	−0.009*x*	−0.032*x*
Discount factor for 10% pa	0.91	0.83	0.75	0.68	0.62	0.56	0.51	0.47	0.42	0.39	0.35	0.32	0.29	0.26	

returns from a particular area of research expenditure will be. They will, however, provide a good means of sifting out those projects which could never generate adequate rewards, highlighting those which are obvious potential winners and classifying the border-line cases from a purely financial point of view. Decisions also need to reflect strategic product range, territorial and 'business shape' objectives.

The investment decision also requires some logical process to guide the judgement of whether or not an invention of the required size could be made in the assumed time scale and to assess the internal and external factors which are critical to the chances of this success.

For a more comprehensive view of ways of making such a judgement, readers should explore the literature on technological forecasting.

For the agrochemical industry, plans for investment in research to generate products which will not enter the market for 10 years, nor generate significant returns for 15 years, must make explicit assumptions about the future. They must consider all those technical, commercial and legislative factors (primarily toxicological and environmental standards) which will determine the success or failure of a future product as a profitable investment. Some of these factors and the trends foreseen are discussed in Chapter 10.

PORTFOLIO MANAGEMENT

At any one time, the manager of a portfolio of R & D projects within an agrochemical company will seek to gain as high an aggregate present value as he can from his portfolio, consistent with balancing risks and satisfying other corporate objectives, including requirements for short, medium and long term cash flow and long term business shape. Too much emphasis on the present value and short term cash flow criteria could stifle longer term investment to the detriment of the future shape of the business.

Each project can be viewed in terms of:

— Its strategic contribution to the business; for example, the way it fits with the existing product range, its contribution to plans for particular directions of corporate expansion or fit with priorities for particular territories.
— Its total size, measured by total sales value in a particular year and its net present value.
— Its quality, measured by the percentage margins made in any one year and the internal rate of return.
— Its risk, measured by the maximum cumulative negative cash flow and coloured by the identified areas of technical and commercial uncertainty.
— Its vulnerability to significant changes in the future market environment, reflected in the payback time — the length of time before the cumulative cash flow becomes positive.

The way each of these factors is viewed depends upon:

— the range of potential projects on offer at the time
— the financial and strategic position of the company at the time

— the scope for getting a balance of risk and reward that the management is happy to accept.

A cash-rich company with good cash flow prospects for the next 5 to 10 years will be much more willing to invest in very large but high risk and long payback projects, than a company with a major cash flow shortage projected within the next 5 years. The cash-poor company would much prefer lower risk, rapid payback projects, even if they afforded the chance of much lower eventual rewards. If they attempted the 'long term reward' strategy they could easily collapse on the way in an annual cash flow crisis.

When considering long term projects in the portfolio, the more fundamental the research and the longer the payback time, the more dramatic the discovery has to be and the larger its commercially exploitable potential.

For most commercial organizations it is very difficult to justify large expenditures on fundamental science, because the business financial pressures demand that funds are used in areas where real rates of return are likely to be high and long term commercial opportunities are realistically achievable. Breakthroughs are very difficult to foresee in any industry. The bulk of the more fundamental work has to be carried out in institutions whose financing and objectives are judged by different criteria. For example, academic research is undertaken in an environment where major objectives are teaching, the development of an enquiring scientific mind and the pursuit of knowledge beyond existing bounds. In industrial R & D, the basic research expenditure is therefore often linked to funding of academic research, providing a high gearing on cash compared with that spent internally, coupled with sufficient internal research to maintain a core of scientists sufficiently skilled and knowledgeable in the critical areas of science to 'gatekeep'. These scientists would be sufficiently immersed in the science, and have a good level of awareness of possible practical applications, to recognize opportunities which may arise in academia and support them when appropriate.

PROJECT MONITORING

As projects progress over the years, more and more is known about each facet and better predictions can be made about the future. It can be very instructive for management to maintain a 'project log' which records the profile of each project as seen at each review date. Over the course of years (and each project can take five to ten years from entering development to reaching the market) it is possible to see whether the project is becoming more or less attractive to the company.

Table 17 gives an example of the sort of tracking that can be done. This shows, for each review time, the relative financial attractiveness of the project and the major areas of technical strength and weakness. At each review time it is possible to compare across projects, to gain an impression of the balance of strength and weaknesses within the portfolio. It is also possible to compare between review times within a project to see whether particular areas of concern are becoming resolved or whether they remain risky.

In the example in Table 17, the reviews of the project in 1988 were generally confident. There seemed to be no particular reason to suppose that technical risks

Table 17 — Example of a project progress log

	Year of Review						
	1988	1989	1990	1991	1992	1993	1994
Financial estimate							
NPV @ 15% £m	20	25	23	12	10	15	20
£m Sales at max	50	60	50	25	20	22	24
% GM at max	70	70	60	60	60	65	60
Year of 1st sales	1992	1992	1993	1993	1994	1994	Sales
Technical risk							
Biological performance	M	M	H	H	M	L	L
Toxicology	M	M	M	L	L	L	L
Environmental profile	M	M	M	M	L	L	L
Manufacturing process	M	H	H	M	H	L	L
Formulation	L	L	H	H	M	L	L
Commercial profile							
Comparative advantage							
— Product	H	H	H	H	M	L	L
— Corporate	M	M	M	M	M	M	M

H, M and L are subjective assessments of high, medium or low risk.
NPV = Net present value.
GM = Gross margin (sales−variable costs of production).

were greater or less than usual and, from past experience, the formulation development area look particularly straightforward. In 1989 views of the market opportunity were more encouraging. A major problem had arisen in manufacturing process development, but it was believed to be soluble soon enough not to delay first sales. In 1990 a major problem had also arisen with the formulation and major biological problems had arisen connected with phytotoxicity. It was not known whether these two events were connected or whether the biological problems were connected with very different weather conditions or particular crop varieties. The concern was such, however, that the first sales date estimate was set back by one year. In 1991 the problems still persisted unsolved, suggesting that the product would be suitable for only about half the originally forecast market. Production processes looked more hopeful however. By 1992 the biological and formulation problems seemed to have been overcome to some extent, but new process problems had arisen, delaying the construction of the manufacturing plant and the date of first sales. In 1993 everything fortunately became sorted out ready for launch and sales in 1994.

PLANNING R & D

The efficiency and the effectiveness of an R & D organization is dependent on both doing the right projects, as guided by the financial and technical risk assessments, and doing those projects in the right way. To be successful, an organization must be both efficient and effective. Efficiency implies a high level of planning and organization to ensure that tasks are performed in ways which use resources most economically. Effectiveness implies a measure of flair and inspiration in seeing imaginative, good, new solutions to problems and generating those all-important insights which crack otherwise daunting problems, and provide effective 'short cuts'.

For the 'near market' projects which are concerned either with expanding and maintaining existing products or with developing and registering conventional and characterized 'development project', efficiency can be sought through:

— clear definition of project objectives
— clear critical path networks
— well-organized division of labour to achieve tasks
— good mechanisms for relating resource costs of particular programmes of work to the financial rewards to be gained
— explicit review and monitoring processes to check on progress to make adjustments to work programmes in response to new information.

For example, a project objective can be set to develop and register a chosen compound for powdery mildew control on vines in France. This can then be broken down into a series of clear sub-objectives, for example:

— assemble toxicology dossier,
— assemble environmental impact dossier,
— define application rate and timing,
— establish optimum formulation,
— estalish commercial manufacturing process.

Each of these can be broken back further into sub-objectives. The objective to assemble the environmental impact dossier can be sub-divided into, amongst other things:

— assemble data on metabolism in vines,
— assemble data on residues of parent compound and metabolites in grapes and wine.

Each of these can be further sub-divided in individual tasks to be performed at particular times by particular people as indicated by the network plan. For example the activity to 'assemble data on metabolism in vines' can be split into:

— synthesize C_{14}-labelled compound,
— formulate compound into typical spray material,
— treat vines with six applications of 100g ai/ha at 14-day intervals,
— harvest grapes,
— generate chromatographic data on metabolites,
— determine identity of metabolites,
— prepare metabolism study report,
— approve report through quality assurance.

Each of these activities can be planned on a network and linked to the other activities with which it has interdependencies. The implications of deviations from plans can be easily seen and requisite actions identified. The costs of each of these tasks in resource and cash terms can be quantified and their expenditure judged against alternative uses.

For the longer term more exploratory or inventive activities, it is less possible to follow such a structured 'production engineering' approach. Objectives are more loosely set because it is not clear exactly what sort of invention will be made or when. The crucial management need is to create the physical and human environment within which an invention is most likely to be made and where attitudes are sufficiently tuned to recognize a worthwhile invention or lead, should it appear. To balance this, management must also have the ability to recognize when a project really is unlikely to succeed and have the courage to discontinue it at the right time!

In the 'conventional' pesticide invention process, this means:

— A concept in the minds of the synthetic chemists and screening biologists of the broad specifications of a worthwhile target invention, for example an insecticide of novel chemical type and mode of action to control Lepidoptera.
— A range of creative approaches to lead to chemical ideas, which may include random testing of chemicals, analogues of known natural or synthetic products, computer modelling of the biochemistry of known target receptor sites in the pest or any other possible source of inspiration.
— A process for following leads as effectively as possible through modelling the relationships between biological activity and molecular structure. This was discussed in more detail in Chapter 4.
— A screening process which can give rapid, cheap and accurate information on whether each molecule tested does or does not possess worthwhile biological properties.

Although it is possible to review performance of the inventive process in terms of numbers of molecules synthesized and the quality of their biological effects, it is usually extremely difficult to predict the chance of a worthwhile invention from the quality of historical performance. Management will tend, however, to concentrate its resources on those areas which seem to show the greatest signs of progress, balanced against perceptions of the possible value of a 'winner', should it appear. Progress can be assessed in terms of whether a previously agreed set of objectives have, or have not, been met.

IMPORTANCE OF FINANCIAL ANALYSIS AND PLANNING

Traditionally, much of scientific endeavour has not been subjected to the interference of economists, accountants and planners. Science has become, however, an increasingly important and integral part of many of the more rapidly growing and profitable industries. Investment in science within the industrial corporate structure has become as important as investment in capital assets for production, working capital, marketing or any other facet of business. In the more 'science-intensive' industries, the proportion of the corporate earning spent on science can be as high as on many of the other areas which have been traditionally subjected to investment appraisal and scrutiny on the basis of financial performance criteria. These organizations cannot now afford to exclude the science expenditure from the same rigour.

The financially successful organizations, as the agrochemical industry matures in the ways described in Chapter 9, are likely to be those which achieve good financial and planning controls in ways which do not stifle creativity and effectiveness. This is something which is much easier said than done!

9

The anatomy of the agricultural chemicals industry

THE BEGINNING

Chemical companies first became involved with crop production technology through the large scale manufacture of synthetic fertilizers, through the production of simple pesticidal compounds or though the production of formulation additives (essentially soaps and detergents) for the crop spraying business. For example, I. G. Farben in Germany and ICI in the UK became committed to agrochemicals through large scale fertilizer production, notably that based on ammonia produced from nitrogen and hydrogen by the Harber process. Small chemical companies set up businesses to manufacture and sell simple inorganic compounds such as lead arsenate. Farmers would prefer to buy a finished product rather than follow a 'do-it-yourself' recipe such as 'dissolve 1 oz of arsenate of soda in warm water, add 16 gallons of rain water and then a solution containing 3 oz of lead acetate'. Other companies sold other simple inorganic substances such as sulphur, copper sulphate or Paris green for pest control.

From these simple beginnings the agrochemical business has grown in value (Fig. 27) and has become transformed into a high technology industry relying upon the invention, development and sale of new 'biological effects' chemicals to fuel its continued growth.

The skills of organic chemistry, which began to be applied by dyestuffs companies in the early part of the twentieth century, led to the development of the first organic agrochemicals; the organomercury seed treatments for cereals laid the foundations of the companies which dominate the industry today. As an example, Bayer, which as an aniline dye producer in the early part of the century, developed the first organomercury seed treatment has subsequently grown to be one of the largest agrochemical producers. Major organic pesticides did not reach the market or make an impact on either agriculture or agrochemical industry growth until after World War II.

The industry has evolved to its current size and complexity through a highly

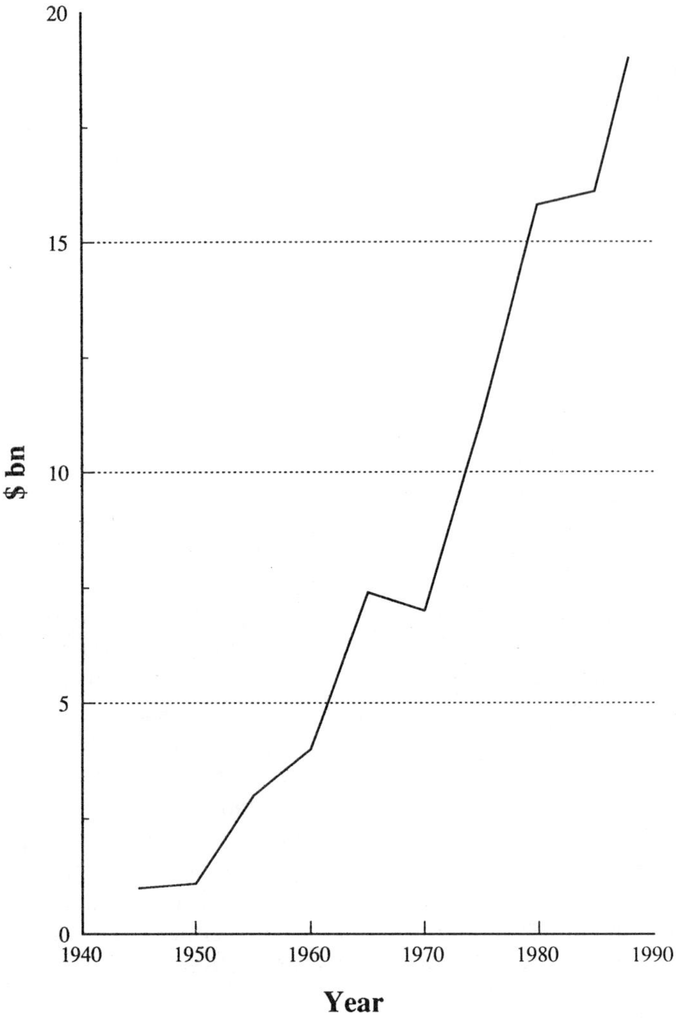

Fig. 27 — World market for crop protection chemicals ($bn of 1986).

innovative and productive period which began in the 1950s. At that time there was a very wide range of unsolved pest problems which caused severe economic loss to agriculture in both yield and crop quality. Those companies which discovered and developed chemicals which could solve one or more of the most important of those problems were able to grow rapidly, creating new markets at a rate determined largely by the speed of adoption of the new technology by their customers. Competition between companies accelerated the expansion of the total market, rather than leading to strong intercompany competition for market share.

The period of most rapid growth for the industry came during the 1960s and 1970s (Fig. 27). Herbicides which had been introduced into the major arable crops of North

America, West Europe and Japan during the 1950s and early 1960s were passing through the most rapid phase of their adoption by farmers. Systemic fungicides first became available which could produce very significant yield increases in wheat and barley in Western Europe. The use of fungicides became a major part of a programme to intensify production of these highly profitable, large area crops with prices which were supported by the European Common Agriculture Policy.

The increased use of chemicals on the major acreage crops dominated this rapid growth phase of the industry, and these same crops will continue to dominate its future. Fig. 28 shows the share of the market, by value, accounted for by the major

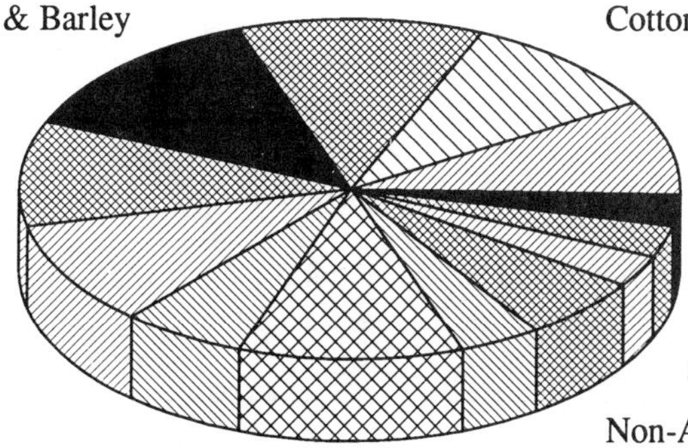

Fig. 28 — Share of the world agrochemical market by crop.

arable crops, wheat and barley, maize, rice, soybeans and cotton, and the major perennial crop, grapes. Fruit and vegetables in aggregate account for large sales, but the total is made up from a very large number of different, individual crops. With the exception of early sales of chemicals into plantation crops, the international market for agrochemicals has been dominated by the high value, technically sophisticated agriculture of North America, Western Europe and Japan (Fig. 29).

Much of the early growth was driven by a relatively small range of new chemicals and many companies grew on the basis of one major discovery, possibly supplemented by a number of smaller ones. Major examples of corporate growth driven by single major products during this period are shown in Table 18.

As the companies exploited their early inventions and corporate incomes and profits grew, there was an increasing ability to re-invest the proceeds in R & D to find and exploit further products. The precedent set by the success of the early inventions and the increasing awareness of the potential for more and better products, provided the incentive for this reinvestment.

The trend in total industry R & D expenditure and expenditure as a proportion of sales is illustrated in Fig. 30 showing strong increases in both parameters. During the

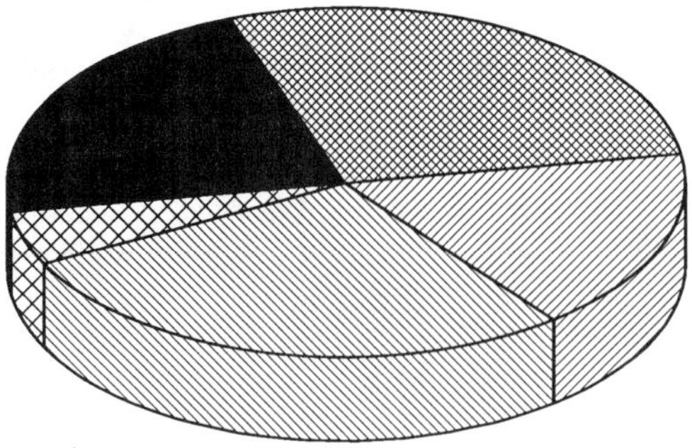

Fig. 29 — Share of the world agrochemcial market by territory.

Table 18 — Major product fuelling early market growth

Company	Product	Introduced	Use
Bayer	Parathion	(1947)	Broad-spectrum insecticde
Eli Lilly	Trifluralin	(1960)	Cotton and soybean herbicide
Ciba Geigy	Atrazine	(1957)	Maize herbicide
ICI	Paraquat	(1958)	Non-residual herbicide
Monsanto	Triallate	(1960)	Wild oat control herbicide
Monsanto	Alachlor	(1966)	Maize herbicide

1970s and 1980s this commitment bore fruit and the range of products available to the farmer grew.

Between 1950 and 1975 new agrochemical products were developed, launched and became established parts of agricultural practice to control most of the major known categories of weed, pest and disease on each of the world's major crops. During this time however, a small number of chemical products dominated each sector of the market (controlling particular pests on particular crops) and competitive pressures from alternatives were relatively small.

INCREASING COMPETITION

In the developed agriculture of the USA, Western Europe and Japan, agrochemical use had become accepted practice on most of the cropped area by the early 1980s and products had become available to cure most major pest problems. Farmers were now being given a choice of products to control their pest problems. For the chemical

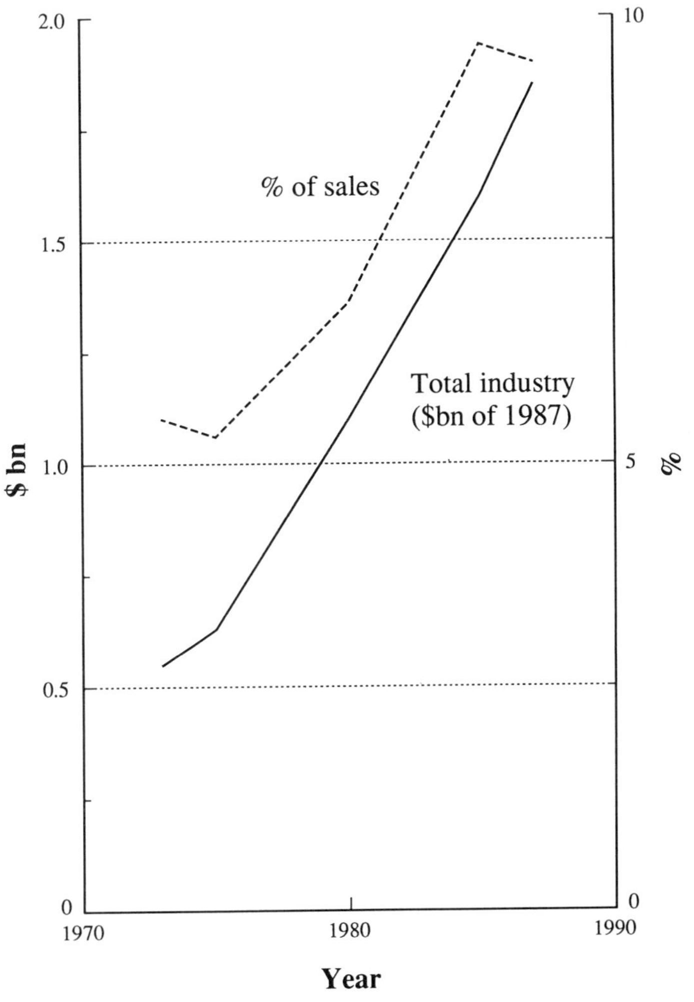

Fig. 30 — Growth of R & D expenditure in the agrochemicals industry, — $bn of 1987, -- % of sales.

companies, the market had become more competitive and corporate growth would have to be achieved more through competition and less from overall expansion of the market.

New product development began to pay increasing attention to advantages compared with existing products. During the early 'competition' phase, most of that advantage has been gained through better biological performance such as better total levels of pest control, covering a wider spectrum of pests or giving a more reliable performance. In certain situations pest resistance to earlier chemical types provided the opening for new chemistry. For example resistance to organochlorine and organophosphorus insecticides provided opportunities for the development of pyrethroids.

As more products have been developed and marketed, the standard of pest control has improved and the technical differences between products, from the users' point of view, has diminished.

The proliferation of 'me-too' products can be seen in many market sectors including triazole fungicides, pyrethroid insecticides and diphenyl-ether herbicides. These products have been developed by companies following parallel research towards similar targets in an environment where each company monitors closely the patenting activities of its competitors. Although there are differences of performance detail between analogous products from different manufacturers, the effects are broadly similar.

Once patents have expired on a number of the major original organic agrochemicals, there has also been a rise in generic production, allowing the same active ingredient to enter the market through different routes from a range of manufacturers as different branded products.

The increasing competitive activity has resulted in progressively better products at lower real prices.

THE CONTRIBUTION OF 'SECTOR CREATION' TO INDUSTRY GROWTH

The overall increase in value of the international agrochemicals business is the sum total of many individual growth areas. These represent:

— the adoption of the first range of products into each crop in the developed agriculture of North America, West Europe and Japan (predominantly between 1955 and 1975);
— the replacement of the original products with better, higher priced products in new sectors of those markets (a major factor from 1975 to the 1990s);
— the expansion of crop protection technology beyond the bounds of the developed world (a major aspect of the likely growth of the business in the 1990s).

The types of technical advances which have created new market sectors were illustrated in Chapter 3. Marketing policies of the various chemical companies have then used the technical differences and their perceived value to the farmer. They have sought to position each product in the market with its own product image and price structure. The product suppliers will make judgements about the relationship between relative prices of their product and others in relation to market share and volume. Farmers will, in time, weigh up relative prices and perceived product advantages in making their purchasing decisions.

The patterns of decision-making may be complex. A farmer may be prepared to pay a premium price for a product which controls a wide range of weeds or diseases either because he does not know which pests to expect, or because he knows he will have a wide range of problems to contend with. Other farmers may choose to buy cheaper, narrower-spectrum products either because they have more specific targets for control or because they do not believe the extra quality or breadth of control is financially justified.

As the number of products available to the farmer has increased, the technical differences have decreased and commercial competition has become more intense.

Increasing emphasis has been placed on price and promotional means of selling; the major tools of competition for maturing industries.

As the standard of products available in the market has risen, it has become increasingly difficult to invent a new product which would constitute a sufficiently significant advance to capture a large market share. Price and other competitive pressures have affected margins and profitability. At the same time, costs of developing new products and sustaining existing ones have risen as a result of increased political awareness of environmental issues and demands for more extensive and detailed product testing.

INDUSTRIAL CONSOLIDATION

These factors, taken together, have brought the industry into a phase of consolidation, in which the structure of the industry is changing rapidly. There used to be many small to medium size companies, each of which established itself after making an invention of sufficient value to fuel its initial growth and many of these businesses were initially able to depend on their own national markets, albeit often augmented by export sales. Costs of development and the need for economy of scale are now such that businesses must be international to survive. As a result the industry has now become increasingly dominated by a few large companies operating on a truly international basis. The survival, size, growth and importance of the companies concerned reflect a combination of good fortune and ability to build upon early successes so as to maintain their momentum of growth as the business has matured. It is only through economies of scale that companies can continue capital investment, fund R & D and grow profitably. The dominance of the large corporation is increasing further through merger and acquisition. Between 1972 and 1988, the share of the world market held by the top 10 companies increased from 57% to 75% and this trend continued during 1989.

CORPORATE STRUCTURE

The corporate structure of the industry is now dominated by the world's largest chemical companies, for whom agrochemicals generally account for between 4 and 18% of total corporate sales by value (Table 19). About 50% of the world agrochemical market at ex-manufacture level is held by the top six corporations and 80% by the top 12.

The largest companies grew from bases in Western Europe (notably West Germany, UK, France and Switzerland) and USA from origins in chemical companies founded in the late 1800s or early 1900s.

By way of illustration, Bayer developed from an aniline dyestuffs company founded in Germany in 1863 which discovered its first synthetic insecticide in 1892. It became part of I. G. Farben Industries AG in 1925 and entered crop protection chemistry as a significant business with the synthesis in 1936 of amidocyano ester of phosphoric acid. After the Second World War the company was re-established as an

Table 19 — Major companies' 1988 agrochemicals turnover

Company	Agrochemical sales ($m)	Total group sales ($m)	Agrochemicals as % of total
Ciba Geigy (1)[a]	2070	11690	17.7
ICI (3)	1910	21041	9.1
Bayer (2)	1870	22735	8.2
Rhone Poulenc (4)	1688	10740	15.7
Du Pont (5)	1412	32360	4.4
Monsanto (6)	1377	8293	16.6
BASF (8)	1025	24645	4.2
Shell (7)	995	79142	1.3
Dow (10)	959	16682	5.7
Hoechst (9)	935	23013	4.1
Schering (11)	740	2962	25.0
American Cyanamid (13)	690	4592	15.0
Sandoz (12)	586	6725	8.7
Eli Lilly (15)	462	4070	11.4
Kumiai (14)	404	429	94.2
FMC (17)	390	3287	11.9
Sankyo (16)	374	2889	12.9
Nihon Nohyaku (18)	355	382	92.9
Rohm & Haas (20)	339	2535	13.4
Hokko (19)	306	344	89.0
Takeda (21)	292	4569	6.4
Sumitomo (22)	276	7147	3.9
Uniroyal (26)	235	734	32.0
Dr Maag (25)	205	5757	3.6
Nippon Soda (23)	200	670	29.9
Nissan (24)	191	681	28.0
Makhteshim (29)	189	228	82.9
Fermenta (28)	172	451	38.1
Pennwalt (33)	148	1024	14.5
Chevron (27)	137	27722	0.5
Nippon Kayaku (30)	124	838	14.8
Mitsui Toatsu (31)	122	3082	4.0
Ishihara (32)	117	546	21.4
Nufarm (—)	104	107	97.2
Griffin (36)	86	86	100.0
Agrolinz (34)	79	367	21.5
Cheminova (35)	75	189	39.7
Agrimont (37)	66	10739	0.6
MSD (39)	47	5940	0.8
Hodogaya (38)	45	261	17.2

Sources: Published accounts and County NatWest Wood Mac estimates.
[a]Figures parentheses indicate 1987 rankings.

entity and grew rapidly with the parathion insecticides launched in 1947 and other organophosphorus insecticides launched in the 1950s.

Ciba-Geigy was formed in 1970 from the merger of the Swiss companies J. R. Geigy and Ciba. Both parent companies also had their origins in dyestuffs, J. R. Geigy dating from 1758 and Ciba from 1856.

ICI was formed in 1926 by the merger of the British companies, Nobel Industries Ltd, Brunner Mond & Co Ltd, British Dyestuffs Corporation and United Alkali Company Ltd. Agrochemical sales initially grew with the development of BHC and MCPA in the UK and Commonwealth.

Du Pont was founded in the USA in 1802 to manufacture gunpowder. Vigorous expansion and diversification between 1918 and 1930 financed by profits from the 1914–18 war led to agrochemical production with seed disinfectants in 1928.

Each of the major companies has grown in size and diversity of chemical manufacture and has become much more international. Because of the international diversity of agriculture, the wide range of crops, climatic zones and pest targets, the agrochemical divisions of those companies are often among the most international within their parent organization. Although their corporate headquarters, and often their principal manufacturing capacity and research base, are still predominantly in their countries of origin, all have partly or wholly owned operating subsidiaries in many countries of the world.

In some larger territories, the national subsidiaries have manufacturing capacity for active ingredients which may supply part of the international as well as local demand. In many countries there are formulation and packaging facilities. In some there are satellite research facilities and, in most, there are field development teams capable of carrying out detailed efficacy and environmental tests.

Without such capacity to carry out local biological testing, it is difficult for a company to establish the correct recommendations for local use of a product. In many territories government authorities insist on local biological efficacy and crop residue data being included with other internationally applicable data in the formal registration submission.

Most of the medium and smaller size companies (generally those listed in Table 20 with sales of less that $500m) do not have the necessary capability to develop and distribute products worldwide and they rely very much on development and distribution agreements with the major companies to exploit their products.

As it is increasingly difficult to sustain a critical mass of R & D without an international turnover of more than $500m, many of the smaller companies are under threat in the medium to long term; indeed many of the smaller western companies have already been absorbed by larger corporations in a steady process of industry rationalization.

Most of the medium to smaller size companies now operating are Japanese, many of whom are relatively recent entrants to the industry and draw profits from an exceptionally highly priced local market and strong non-agricultural businesses. They seek to make an invention of sufficient international importance to enable them to expand into the truly international league.

Many of the Japanese companies have established strong research, development and marketing links with the international majors and some have formed formal joint ventures with one or other of the majors.

THE GROWTH OF LEGISLATION

The corporate shape of the industry is also being moulded by increasing legislative pressure. Governments have, from the earliest days of the industry, played a role in promoting the discovery and testing of new chemicals. As the commercial chemical industry became fully established the rôle of governments evolved into that of ensuring that products are effectively and safely employed. This has inevitably entailed the development of a complex legislative framework which governs the registration and sale of all agrochemical products, an issue discussed in Chapter 5.

Probably the first legislation affecting the industry was the 1910 United States Federal Insecticide and Fungicide Act which was enacted to control the many 'quack' remedies being launched onto the market at that time. Manufacturers were forced to state on the package label the nature and content of the active ingredient and had to be able to substantiate their claims of biological utility.

After the 1914–18 war the German government introduced a voluntary testing scheme to enable manufacturers to register their products, the list of organisms controlled and the rate of application required. For products to gain registration they had to be proven effective in a well-organized series of trials extending over three years at official German agricultural research stations. This proved invaluable to both farmers and the German agrochemicals industry and was the forerunner of the German official testing scheme that exists today. Many other countries followed suit and introduced testing schemes.

As the century advanced, however, legislation progressively reflected increasing concern for safety and the environment (discussed in Chapter 5), and it is these concerns, rather than those for biological performance, which have resulted in an ever-increasing array of national and international legislation. During the latter part of the 1980s when food production in the developed world tended to exceed demand, political pressures in support of food production and agriculture lessened. The countryside became increasingly valued as an amenity rather than as a factory for food production, and an urban society became far more concerned about the environment and human health. Concerns encompassed such matters as the reduction in occurrence even of minute levels of residues in food, obviation of risk of pollution of underground water, and protection of 'non-target' flora and fauna.

The legislative pressures have increased the demands on companies for comprehensive programmes of product defence, and necessitated complex programmes of regulatory studies for new product registration. Smaller companies have found it increasingly difficult to provide the resources required to match these demands and to field the necessary level of skill and expertise in negotiation with many independent governments. These pressures, therefore, have led to further company consolidation.

THE FUTURE

The agrochemicals industry will become increasingly concentrated in a few very professional, well-organized, environmentally concerned, international corporations which will have greater proportions of their businesses in the developing agriculture beyond the borders of the 'home markets' of North America, Western Europe and Japan.

The next chapter speculates on some of the new influences which may shape the industry over the next quarter century.

10

The future

PRESSURES WHICH SHAPE THE FUTURE

During the second half of the twentieth century world agriculture and the scientific nature of its technical base have changed dramatically. In the developed countries, populations have become more urban and affluent. Political attitudes to agriculture, the countryside and environmental issues have reflected this change. In the developing world, population numbers have continued to escalate while scope for expansion of agricultural areas has been limited. This has generated enormous pressure to intensify production. As part of the process of intensification, the agrochemicals industry has grown from almost nothing to become a vital, highly technical but fully integrated part of the agricultural industry.

The pace of change is not decreasing. We can only speculate on the shape of the industry of the early twenty-first century, but the major pressures which mould it derive from:

- population expansion and demands for food quality;
- economic growth and increase in income per head, driving demands for food quality;
- new developments outside agriculture which create demand for agricultural products as industrial feedstocks;
- the needs of agriculture for new or better technologies to improve productivity and sustain farm incomes.;
- the scope for greater and more effective use of existing agrochemical technology;
- advances in science, which create scope for new agrochemicals technology which can, in turn, lead to new markets for agrochemical products;
- advances in science which enable substitution of chemical with non-chemical methods of pest control;
- regulatory pressures from governments and the changing importance of 'environmental politics';

— commercial pressures arising from competition between agrochemical producers and from the competition of 'alternative solutions'.

THE NEED FOR FOOD

The earlier discussions in this book have highlighted the increasing pressure that world population growth is placing on food production. Rising standards of nutrition are vital.

The domestic political changes taking place in Eastern Europe during the last decade of the twentieth century are dependent for their long term stability upon the ability of those nations to feed their populations at an acceptable cost, providing food at prices that people can afford and in volumes which avoid long queues and rationing.

Stability of relationships between rich and poor countries will be partly affected by the ability of the inhabitants of the poor to feed themselves and not create unacceptable political pressures through fundamental needs for survival.

Shipment of food through aid and international trade from the productive 'grain baskets' of Northern America and Western Europe, will markedly influence the size of market for which those 'high technology' agricultures can produce.

Most of the added production will come from increased production per hectare, because of the limited availability of good agricultural land area. A large proportion of that increase must take place in Eastern Europe, where productivity levels are well below those in Western Europe and North America, and in the developing countries where population growth is greatest and nutritional standards are lowest.

Fertilizer usage statistics, as an index of agricultural intensification, show that rapid changes in production technology are taking place in most parts of the world. Rates of change are generally highest where absolute levels of technology are lower in relation to potential.

Agrochemical usage will become a major part of the new set of production inputs in every agricultural system. Under conditions of lush tropical agriculture, weed, insect and fungal problems will become even greater than in the temperate North as fertilizer usage rises and production intensifies.

In the highly developed agriculture of Japan, West Europe and USA, the pattern of change will be very different. Demand for increased total production will grow extremely slowly with very low rates of population growth. Demand for volume of agricultural output will depend upon exports. There is likely to be continued downward pressure on real prices of agricultural products which will force farmers to minimize production costs per tonne. This will be achieved partly through reduced overhead costs derived from economies of scale on larger farms with larger equipment, and fewer man-hours per hectare. It will also be achieved through higher average yields. Cropped areas will tend to retreat from marginal land into those areas most favourable for the particular crop. On these more favoured areas, inputs such as agrochemicals, fertilizers, and improved varieties will be used increasingly intensively, up to the point at which the marginal value of the increased yield is closer to the incremental cost of the marginal inputs. At present the marginal benefit of agrochemicals is often several times bigger than the marginal cost. The real cost of chemicals to the farmer is also falling. Intensive agriculture is likely, therefore, to continue to have high levels of use of technical inputs.

Overall, the range of agricultural pressures on the agrochemical industry is likely to result in:

- An increasing proportion of the value of the world agrochemical market being in the 'developing' countries as agricultural production there becomes more intensive and uses more 'technical' inputs.
- An intensification of chemical use on a declining or static area of crop in the developed world. The rate of intensification is likely to be slower than it has been and, taken together with falling real prices for agrochemicals, is likely to result in a slow growth of total market value. This does not imply an increase in tonnage of pesticide used, as there will be a shift to more active, high quality products.

AGRICULTURAL PRODUCTS AS INDUSTRY FEEDSTOCK

Some crops have always been grown for reasons other than food production. These include cotton, rubber, flax and jute as examples. New manufacturing processes are, however, creating opportunities for other crops. In Brazil, ethanol made by fermenting sugar cane supplies a significant part of the demand for motor spirit. Processes are being invented to use starch as the raw material for bio-degradable plastics. The growth of demand for agricultural products as a renewable industrial feedstock is very hard to predict but it will supplement to some extent that projected for food. It is likely to be driven more by the value to industry or consumers of the properties of the natural product as exhibited in preferences between cotton and synthetic fibres or synthetic and natural rubber, than by uses of biomass as an energy source.

Chemicals or biotechnology may provide ways to enhance particular properties or facilitate the industrial processing of crops.

SCOPE FOR IMPROVEMENTS IN AGROCHEMICALS

The major growth phase of the industry has seen the competitive discovery and development of progressively better products. Quality and spectrum of performance is generally extremely good, and product replacement has become more a matter of cost effectiveness than effectiveness as such.

Scope for product replacement will continue to arise as a result of:

- Resistance of insects, fungi or weeds to the existing biochemical modes of action. There will be a major need eventually for a new group of insecticide toxiphores to replace the pyrethroids, just as pyrethroids have ousted organophosphorus and organochlorine products from many markets. Similarly there will be scope for a new group of systemic fungicides to replace the triazoles, which have in turn superceded the benzimidazoles in many markets.
- Shifts in pest population. The effective control of particular weeds or fungi has tended to leave some ecological niches in which previously less important species are able to flourish. Changes in cropping patterns and the importance of certain

crops such as sunflowers and oil seed rape in Europe also bring with them changes in weed and disease problems to be solved.

— Remaining unsolved problems. The industry has still failed to find truly cost-effective products to control nematodes in soil or cost-effective treatments for soil-borne fungal diseases. Estimates of the size of potential markets for uncontrolled soil-borne problems are very hard to make because, except in those fields where problems are very severe, it is not possible to assess the extent to which yields are generally depressed by root problems. Some estimates of total crop loss due, for example, to the soil fungal disease take-all (*Gaeumannomyces*) are very large, and new developments in bio-diagnostics may make areas of severe disease occurrence easier to identify and more predictable to treat. Increased use of agrochemicals in the developing agricultural economies is also likely to face the industry with new problems of pests not adequately controlled by products developed for the markets of North America, Europe and Japan. Each geographical area and climatic zone favours a different range of species, and the impact of agricultural intensification on pest flora and fauna have yet to be fully appreciated.

REGULATORY PRESSURES

Regulatory pressures from governments are increasing worldwide. During the 1970s and early 1980s, the major concerns centred on mammalian toxicology and safety testing concentrated on work to assess hazards to operators and food consumers. There was increasing emphasis on the need to understand the metabolism of pesticides in plants and to measure the residue levels of parent chemical and metabolites in the harvested crop, processed food products and crops following in rotation.

During the 1980s environmental concerns have come increasingly to the fore. Progressively more sensitive detection methods have enabled analysis of minute residues (0.1 parts per billion) in water. Environmentally conscious political lobbies have pressed through legislation seeking to curb agrochemical residue levels to below this in drinking water (see Californian State legislation AB2021 — the Clean Water Act, and EEC legislation). The stringency of the new combination of toxicological and environmental requirements has made the cost of development of a new agrochemical much greater, and the probability of achieving registration much lower. This will reduce the rate of new product introductions and limit development to those new products which are likely to find very major markets in major crops. The anomaly of rejecting new, relatively safe products while allowing continued sales of older, less thoroughly tested products has been tackled by several governments by the introduction of a re-registration process. The US Environmental Protection Agency, for example, has embarked on a programme to re-register all existing products. All studies for existing registered products must at least conform to the 1982 guidelines, and where these have been modified significantly, more recent standards. If they do not, the studies must be repeated and approved within a short, agreed timescale.

The effect of this legislation will be the loss of registration for those products which, on re-test, fail to meet the latest criteria. Companies will also discontinue those compounds whose sales value and profitability are insufficient to justify the costs of re-registration. Some compounds will also be lost because the companies who produce them do not have the funds or facilities to carry out the testing programme necessary for re-registration. It is estimated that about one-third of the crop protection chemicals registered in the USA could be withdrawn as a result of these pressures.

The net effect will be:

— A reduction in the number of agrochemicals available to the farmer, particularly products for the smaller and more specialist crops.
— A further reduction in the number of small companies operating in the industry.
— Greater opportunities for large, well-defended products from large companies having strong toxicological and environmental testing facilities.

Paradoxically, however, the increased attention to the fate of chemicals in the environment and the effect on non-target species will result in opportunities for new products with more favourable environmental profiles to replace existing products.

COMMERCIAL PRESSURES

In common with most other industries, agrochemicals will become increasingly competitive with greater commercial and financial pressures. Each major company will seek to expand at a rate which is more rapid than the total growth of the market. Growth will come, therefore, from increased market share for the successful, resulting in battles for survival for the less successful. The losers will either retrench into market niches where they have some comparative advantage or become absorbed through merger or take-over into larger organizations. The industry is already showing a marked trend to concentrate into fewer and fewer companies.

Those companies which are successful will retain their positions through maximizing the long term cash flow from their established portfolio of products and concentrating research effort towards those new product areas which show the largest potential rewards. Maximizing returns from the established portfolio of products will involve:

— process research and capital expenditure to ensure minimum costs of production, gaining maximum margin, scope for price flexibility and ability to defend against any cheaper generic manufacturers;
— vigorous product defence through generation of new toxicological and environmental data to meet latest requirements;
— development of mixtures, improved formulations and better product presentations to gain commercially competitive edges in the market;
— actively seeking opportunities in the expanding markets created by technical intensification of agriculture in the less developed areas of the world.

New product development will be driven by quality of technical performance as before, but will pay even more attention to cost reduction per hectare of use (through achieving excellent performance at very low rates of application per hectare) and to the search for molecules with better environmental properties than those which currently exist. Soil persistence, movement in soil and safety to beneficial species will all become even more important considerations early in the new compound selection process.

ADVANCES IN SCIENCE WITHIN AGROCHEMICALS

Advances in science within the agrochemicals industry influence both the ease with which new compounds can be discovered and the difficulty of progressing them to registration.

As discussed in Chapter 4, advances in chemistry lead synthesis into previously unexplored fields. The development of biochemical molecular modelling and computer graphics has led to a greater understanding of the physical and chemical properties of molecules needed to optimize biochemical performance in certain sites of action within the target species. Growth of knowledge will also make it more possible to predict the likely environmental properties of a new molecule. These advances help to offset the negative effect of the greater stringency in performance and safety criteria which a compound needs to meet to be worthy of development.

It is probable that the best companies in the industry will continue for a long time to maintain a flow of new products to refresh their product ranges and drive their long term growth.

Advances in science, however, also enable more detailed questions to be asked and answered about toxicological and environmental safety. In a strongly environmental and safety conscious political environment, this means that those questions will be asked. The agrochemical industry is already one of the most regulated. The public concern about 'pesticides', *per se*, also places it under an often highly emotive spotlight. The fact that biologically active compounds are sprayed into the environment leads politicians, the public and regulators to ask the range of questions referred to in Chapter 5 about the fate of those chemicals, and their series of breakdown products in crops, soil, water, rotated crops and processed agricultural products. Advances in the science of analytical chemistry make it possible to trace the fate of much smaller quantities of substances in the environment. Even if those quantities are minute in relation to safety levels indicated by toxicological studies, their mere presence can cause emotive concern. There is often a level of concern which is out of all proportion to the accepted, but not perceived, natural risks or voluntary risks from other aspects of life in the environment.

NON-CHEMICAL PEST CONTROL

Non-chemical solutions to pest control problems have been in existence for some time. The major solutions are the use of natural predators, pesticidal microorganisms and the development of genetic resistance.

Good insect pest management attempts to balance chemical control with the use of natural predators. In those situations where natural predators can play a major role, there has been considerable effort to develop integrated pest management programmes. With knowledge of pest and predator life cycles and ecology plus careful selection and timing of the chemicals used, it is often possible to achieve a more ecologically sound and effective long term pest control strategy than is possible by relying on either chemical or predator exclusively.

These programmes may make use of very selective pesticides or use insect pheromones. An example is the use of sexually attracting pheromones to confuse and disrupt the behaviour of a target insect. This is in practical use, for example, for control of pink bollworm (*Pectinophora gossypii*) in cotton in Egypt.

Greater resistance to pests and diseases exists in abundance in nature. Many pests and diseases tend to be relatively specific to particular host plant species. For example the vine powdery mildew fungus will not infect or grow on any other plant species. To that extent, all other plant species have a built-in resistance to vine powdery mildew. Within a plant species, certain varieties carry greater genetic resistance to disease than do others. The genetic resistance is often due to the ability of the plant to produce natural chemical pesticides. Often that genetic characteristic can be identified and bred into commercial varieties through conventional plant breeding programmes. Genetic fingerprinting is beginning to be used by plant breeders to diagnose the presence of disease resistance genes. This technology will increase the efficacy of plant breeding by increasing the precision of selection and enabling different resistance genes to be accumulated in a single plant variety.

In the future, it will be possible physically to isolate the DNA comprising a resistance gene, and then reintroduce the gene to a non-resistant target plant. This will circumvent the lengthy conventional breeding process. It will also make available a far wider selection of genes than plant breeders at present have access to, since genes can be isolated from species completely unrelated to the target crop.

It is now also possible to incorporate genes for herbicide resistance into crop plants. This enables the use of broad-spectrum weed killers without having to find compounds which show the required selectivity as well as an acceptable spectrum of weed control. For example genetic resistance to the broad-spectrum herbicides glyphosate and glufosinate-ammonium has been demonstrated.

In the long term, there are likely to be many ways in which advances in biotechnology could generate new ‘biological effect products’ for agriculture, thus entering and augmenting the market developed with synthetic agrochemicals during the twentieth century. Genetic effects could supplant or supplement chemical effects. As we saw in earlier chapters, many of the first agrochemicals were natural products, including insecticides made from steeping tobacco leaves, or from pyrethrum flowers. The insecticidal properties of *Bacillus thuringiensis* (BT) were discovered in the early 1900s, although its use has been relatively limited because of its slow activity, allowing plant damage from feeding insects while the toxin takes effect.

Areas of possible future development include:

- Microbial products which could produce crop protection effects directly, as does BT.
- Microorganisms which carry genes for the production of crop protection toxins within plants.
- Modified plants with
  - greater genetic resistance to pests and disease;
  - genetic mechanisms to interact with microbes, for example enhance interac-

tion with nitorgen-fixing bacteria or ability of new species such as wheat to fix nitrogen;
  - improved quality attributes such as controlled ripening or flavour development in fruit through modulation of biochemical pathways. ICI Seeds have produced a modified tomato with improved flavour and storage properties through regulation of the cell wall degrading enzyme, polygalacturonase;
  - improved properties as industrial raw materials.
- Other effects, for example, using ice-nucleating bacteria to improve the frost tolerance of crop plants.

BIOLOGICAL PESTICIDES

Several companies are now attempting to develop and market biological pesticides. The most advanced products, accounting for the majority of the biological pesticide marketed in the late 1980s, are improved strains of *Bacillus thuringiensis*. The major markets developed so far are those in which the slow speed of action is relatively unimportant but 'environmental acceptability', notably selectivity, carries a premium, especially in forestry, mosquito control and on vegetables in California. It also plays a major role on pests which rapidly become resistant to chemicals, for example *Plutella* sp. in vegetables.

The growth of commercial interest has led to the discovery of novel strains of BT, with groups active on Lepidoptera, Coleoptera and Diptera with a range of breadths of spectrum of species controlled.

Strains of BT have been developed and are being marketed for control of mosquitoes, houseflies, forest pests and a number of pests on field crops such as cotton and soyabean.

Examples of areas of research and development for new biological pest control agents are numerous. Some examples serve to give a flavour of these.

Work to explore the use of beneficial bacteria includes:

- *Pseudomonas* spp. and *Bacillus* spp. to inhibit the growth of root pathogenic fungi, allowing better root development and yield, for example non-pathogenic *Agrobacterium radiobacter* marketed by ICI Australia to control crown gall (*Agrobacterium tumefaciens*) on young fruit trees.
- Bacterial organisms to control nematodes.

Viruses have been developed for the control of insects such as nuclear polyhedrosis viruses for control of the forest pests gypsy moth and Douglas fir tussock moth.

Research to explore use of fungal toxins (mycotoxins) to control other pests include products to:

- Protect timber from other fungal pathogens.
- Control insect pests. For example the fungus *Entomophaga gryllii* produces substances toxic to grasshoppers.

As fungi sometimes kill plants through the production of toxins, there is possible scope to explore and expoit these fungal toxins as herbicides. It may be possible to exploit the different natural pathenogenicity of fungi to different plants to achieve selectivity of herbicide action. A product based on the fungus *Colletotricum gleosporoides,* for example, has been demonstrated to control northern joint vetch, a weed in rice.

Nematodes which are natural insect pathogens are being explored as possible means of controlling insects. The more virulent natural pathogenic nematodes exist in symbiosis with a bacterium *Xenorhabdus* sp which is lethal to infected insects.

NITROGEN FIXATION

There is a small market, notably in the USA and Italy (*c* £20m in the USA in the late 1980s), for rhizobia soil and seed inoculants to enhance nodulation and nitrogen fixation by the major legume crops; soyabean, peanuts, alfalfa, beans and peas. There may be scope for advances in producing new strains of rhizobia which fix nitrogen more efficiently than natural strains or which colonize legume roots more effectively than wild strains.

POSSIBLE IMPACT OF GENE TECHNOLOGY ON 'BIO-PESTICIDES'

Many of the first products to be derived from biological sources grown in culture media are likely to derive from natural organisms or improved strains of organisms produced by conventional mutation and selection processes. In the future further advances will be made through the application of newer genetic technologies such as the use of recombinant DNA.

Recombinant DNA techniques allow DNA from one organism to be removed and combined with the DNA of another, transmitting characteristics from the donor to the recipient. The new host organism reproduces the modified DNA as though it were part of its natural genetic material, and expresses the characteristics caused by the protein production for which the new DNA is responsible.

The most widely researched 'bio-pesticide' target is the insecticidally active bacterium *Bacillus thuringiensis*. This is attractive as a first target because its genetic structure is fairly well known and its toxin production is controlled by between one and five genes.

Targets for gene modification would be to produce variants of BT which have greater agricultural utility, for example, a better spectrum of control, speed of action or duration of control.

DNA technology has been used to insert the gene responsible for toxin production from BT into the bacterium *Pseudomonas fluorescens*. The bacterium, which occurs naturally in soil, lives in association with plant roots. The hope was that the bacterium containing and expressing the BT toxin genes would colonize plant roots and protect against soil pests, particularly cutworms. BT genes have also been cloned into *Clavibacterium xylii,* a natural endoplyte of Bermuda grass, which is being tested as a pressure injection into maize seeds for subsequent control of corn borers.

Research is also progressing to introduce the 'BT toxin genes' into crop plants, enabling them to produce endogenous insecticides. These new internally produced substances would then augment the natural endogenous insect control substances that the plant produces from its natural genetic inheritance.

PRODUCTION TECHNOLOGY FOR BIO-PESTICIDES

Successful large scale production of bio-pesticides will require both the discovery and development of organisms which produce toxins giving the required agriculturally useful effects in the field, and the development of suitable large scale production techniques. The most advanced products at present are those which can be made using relatively simple fermentation technology.

TOXICOLOGICAL AND ENVIRONMENTAL ATTITUDES TO BIO-PESTICIDES

In the USA and in many other countries, legislation is being enacted to regulate biotechnology and the use of products of biological origin. As legislation to control synthetic pesticides has grown with time, experience and understanding, so will that to regulate bio-products.

If live organisms are to be sprayed, their ability to grow, replicate and distribute their genetic material gives rise to a new set of questions about environmental impact. In the USA, the EPA requires, for any organism which is not natural or indigenous to the area, information on its identity, growth, survival and competitive characteristics, its host range and potential effects on non-target species, its parental strains and the nature of any genetic manipulation. Large scale field testing in the USA requires an experimental use permit (EUP) in the same way as a synthetic pesticide does.

The long term environmental impact of the new technologies is not, however, clear. Ice-minus bacteria intended to colonize the surface of plants to stop ice forming (thereby increasing frost tolerance) have been produced using recombinant DNA techniques to remove the gene sequence responsible for generating ice-nucleating proteins. These organisms displace the natural ice-plus forms on the leaf surface. It is not yet known to what extent the ice-minus forms can exchange genetic material with other forms. There were fears expressed that if the total population of ice-plus forms were to be reduced in favour of ice-minus forms, there could be an adverse effect on the role of ice-nucleating bacteria in the formation of rain droplets and rainfall! Strains of ice-plus bacteria (strains of *Psuedomonas syringae*) are, in fact, used to enhance the efficiency of snow-making equipment at ski resorts.

CORPORATE INVOLVEMENT IN BIO-CHEMICALS

Much of the pioneering work to seek and develop substances of biological origin as agricultural effects products has been done by small, new and specialist companies, notably in the USA. These include; Ecogen (founded 1983), Mycogen (founded 1983), Advanced Genetic Sciences, Bio-technica, and Genetics Institute (founded 1980). A number of the larger agrochemical companies are also investigating biological sources as a new, alternative source of toxophores.

This area of business also holds potential interest for companies with expertise or excess capacity in fermentation and nutrient culture technologies, such as Abbott Laboratories and Sandoz who market BT, and Merck & Co who market avermectins, insecticidally active compounds from cultures of selected strains of soil *Strepto-*

myces. Grace have developed a *Trichoderma* species. Gustafson have a *Bacillus subtilis* strain for soil disease control.

In the early flush of enthusiasm for biotechnology, many small companies were able to raise speculative venture capital on the chance of a valuable breakthrough. Mature reflection on rates of progress and evident increases in difficulty of registration of bio-engineered products could make financial pressures on bio-technology companies more stringent. The products most likely to reach the market from a regulatory standpoint are those which are either dead or closest to naturally occurring organisms. These are, however, the least likely to offer the functionality that farmers are seeking as an alternative to an available synthetic agrochemical. The market in general may not give preference to substances of biological origin; judgements will be made primarily on cost-effectiveness. It is perhaps significant that the most successful 'bio-pesticide', BT, is naturally occurring, can be produced cheaply by fermentation, and sells into a specific pest niche in environmentally sensitive situations, e.g. forestry.

BIOTECHNOLOGY TO COMPLEMENT AGROCHEMICALS

A number of the larger agrochemical companies are investing in biotechnology research to develop genetic lines of crop plants resistant to their broad-spectrum herbicides. This would allow an increase in their conventional product application (and sales) in areas where product usage would otherwise have been constrained by phytotoxicity to the crop.

The first patent for a herbicide-tolerant maize plant developed using biotechnology, was granted in 1988 to Molecular Genetics Inc. The patent covers maize plants resistant to imidazolinone herbicides. The seed company, Pioneer, is expected to develop proprietary maize lines containing these genes for sale in the early 1990s.

AN OVERVIEW OF THE FUTURE

Agrochemicals will continue to be a fundamental part of the technology required to feed the burgeoning populations of the world to a higher nutritional standard with cheaper, higher quality food. The use of chemicals will expand as developed agriculture becomes even more intensive and as developing agriculture selectively adopts the technologies of the developed world.

The products used will be tested and approved with even greater thoroughness and products found to have unacceptable safety profiles will be replaced by safer ones. The levels of safety judged to be acceptable will be very high and judgements are likely to remain disproportionately stringent when compared with other sources of personal risk. Advances in science will aid the discovery process and help to balance the apparent reductions in probability of discovery which result from increasingly stringent market and safety standards. Agricultural technologies derived from biotechnology will augment, and to some extent compete with, those of chemical origin but their impact is unlikely to be very significant until the twenty-first century.

The industry will follow the same pattern as that followed by so many other high technology industries and become progressively more concentrated in fewer large international companies, which are forced by intensifying competition to become increasingly effective and efficient.

The future success of the industry will be judged by the extent to which it meets the aspirations of those whose interests it serves. These are the needs of farmers to produce high quality, low price food in a competitive agricultural business; the needs of food consumers for cheap, high quality, safe food; long term care for the environment; and profits for its own growth and regeneration.

Appendix 1: Competitive introduction of chemical analogues

Most crop protection chemicals come from a limited number of chemical 'families'. Most insecticides, for example, belong to major groups such as the organochlorines, organophosphates, carbamates and pyrethroids.

The triazole group of systemic fungicides provides a recent example of a very prolific chemical area where many companies have managed to define groups of molecular structures which are different enough to constitute separable patent property. All 1,2,4-triazole fungicides share a common mode of action, which is to inhibit the action of the 14 α-demethylase enzyme in the biosynthesis of ergosterol in fungi. The 1,2,4-triazole ring is important for blocking this cytochrome P450 enzyme by binding to the iron haem at the active site. The remainder of the structural frameworks have sufficient differences to be distinguishable areas of chemistry for patent purposes, but have sufficient similarity in shape, size and physical properties to be transported within plants and fungal mycelium to the active site and to fit within that site. Each molecule does have its own strengths and weaknesses from the users' point of view, enabling it to be positioned slightly differently in the market.

Examples of the structures of commercial triazole compounds from a number of different companies are set out in the figures. Some of the differences are:

- Diclobutrazol from ICI has a carbon atom in place of the oxygen in the 'backbone' of Bayer's tradimefon.
- Diniconazole from Sumitomo has an unsaturated double bond in its central structure compared with the saturated version in diclobutrazol.
- Propiconazole and etaconazole from Ciba-Geigy and furconazole from Rhône-Poulenc have oxygen-containing heterocyclic rings in place of the open alkyl configurations of the triadimefon and diclobutrazol types.
- The Ciba-Geigy and Rhône-Poulenc structures differ in the number of oxygens in the heterocyclic ring.
- Sandoz's cyproconazole and the ICI compounds flutriafol and hexaconazole differ essentially in the fourth substituent on the central tertiary carbon. Flusilazole from Du Pont follows a similar theme to flutriafol but with the major distinction of having a central silicon in place of the central carbon.

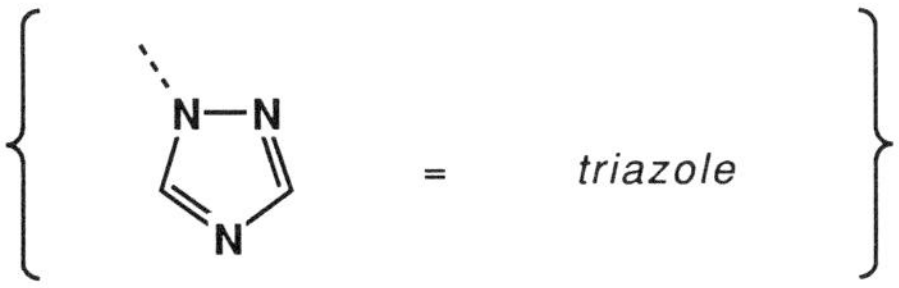

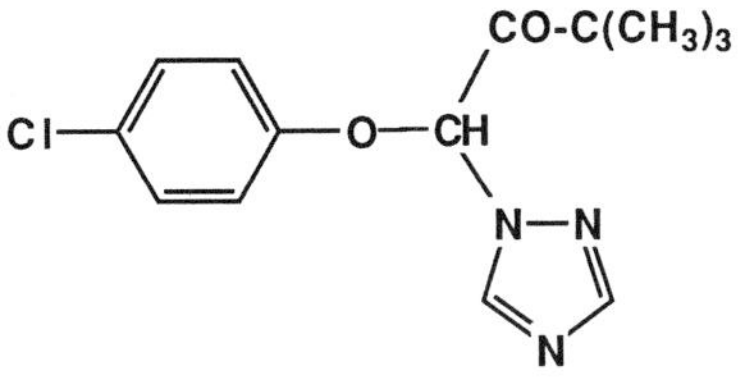

triadimefon
(Bayer, 1973)

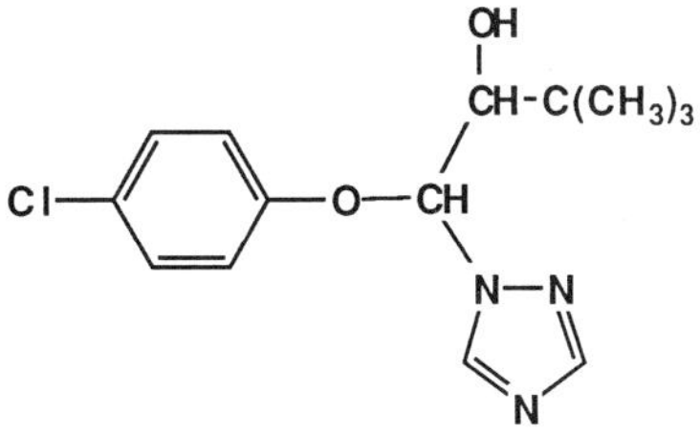

triadimenol
(Bayer, 1978)

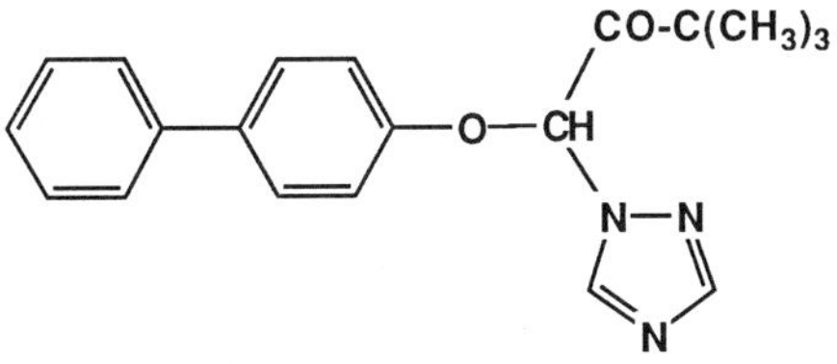

bitertanol
(Bayer, 1979)

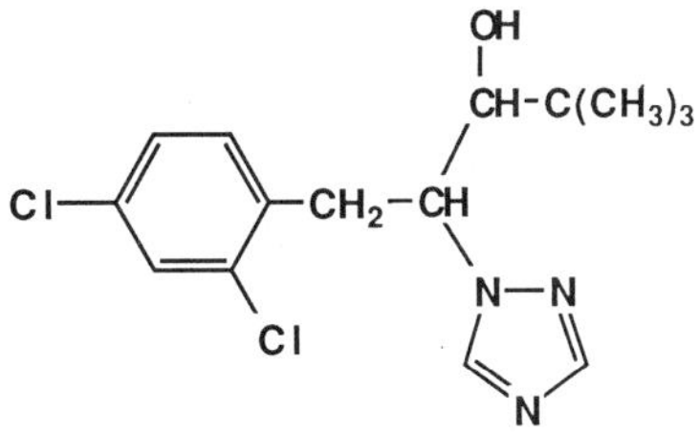

dichlobutrazol
(ICI, 1979)

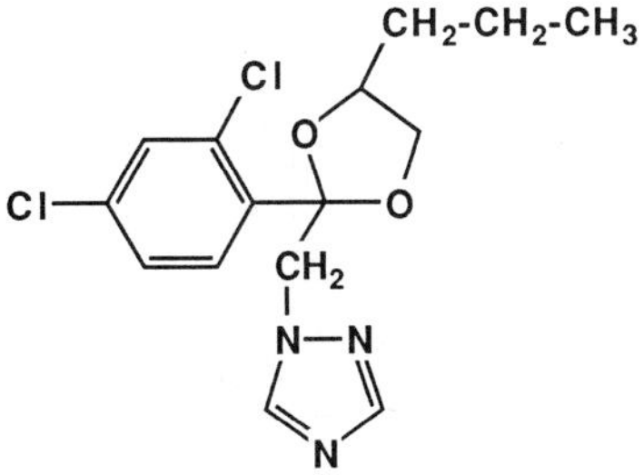

propiconazole
(Ciba-Geigy, 1979)

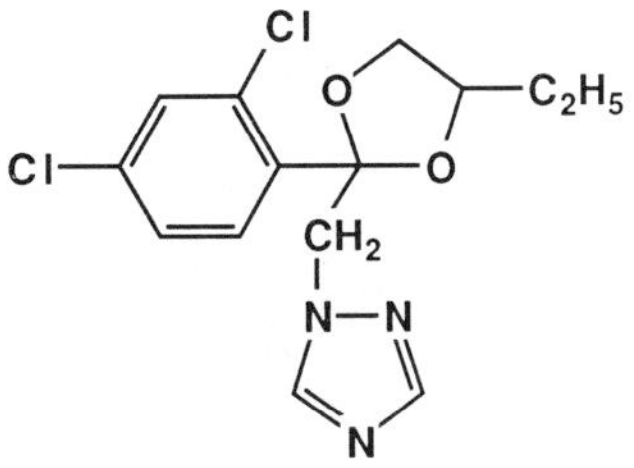

etaconazole
(Ciba-Geigy, 1979)

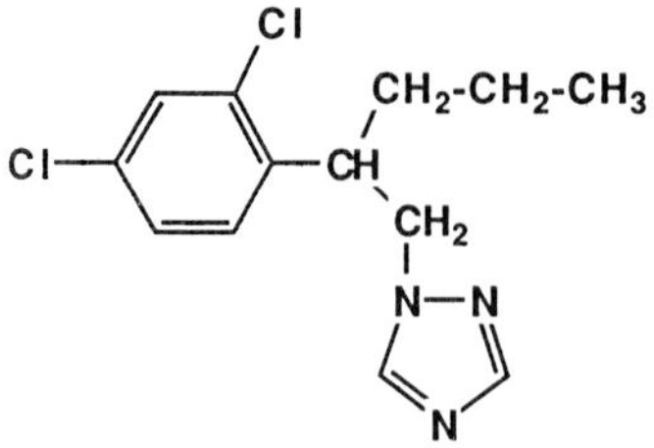

penconazole
(Ciba-Geigy, 1983)

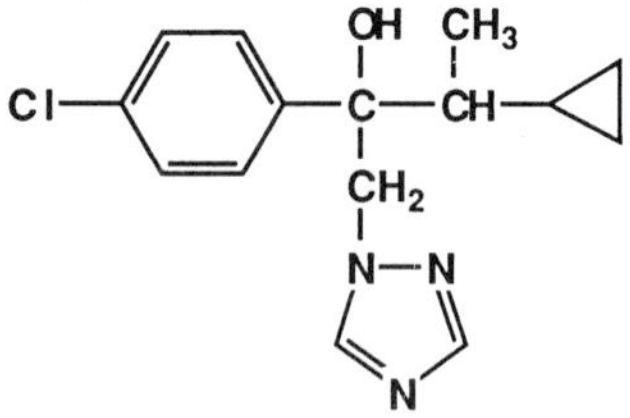

cyproconazole
(Sandoz, 1986)

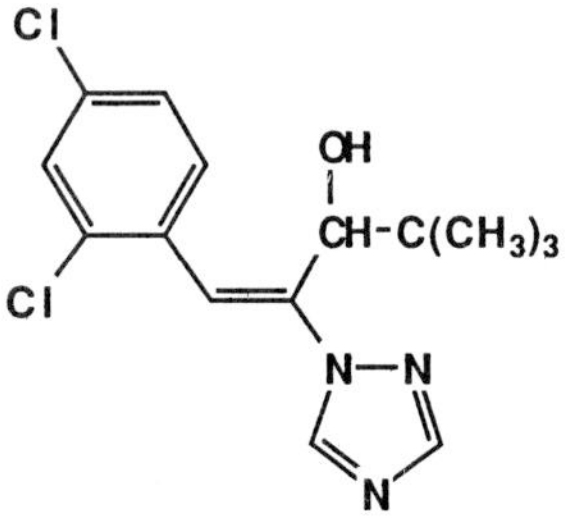

diniconazole
(Sumitomo, 1982)

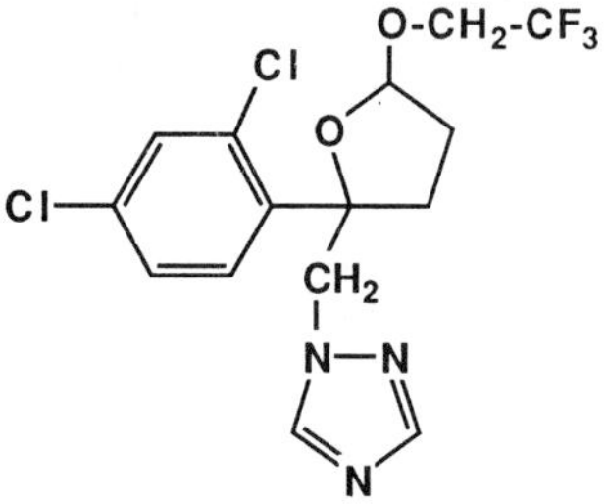

furconazole
(Rhone-Poulenc, 1988)

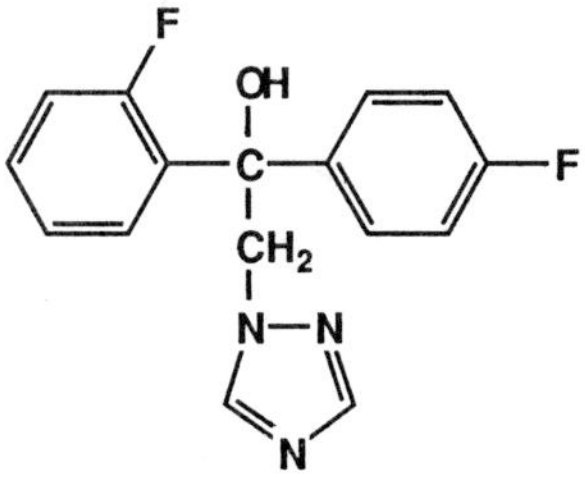

flutriafol
(ICI, 1983)

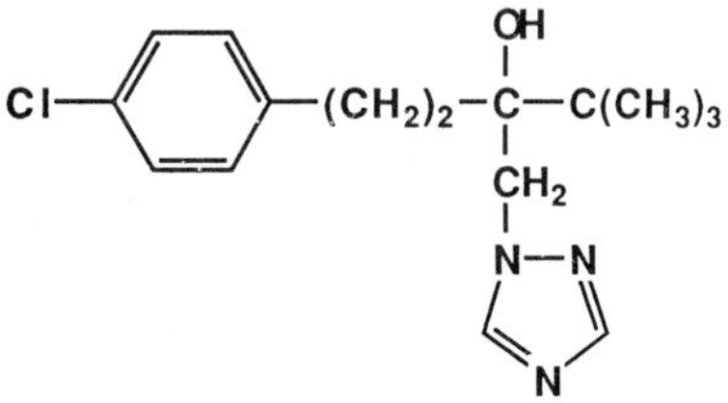

terbuconazole
(Bayer, 1986)

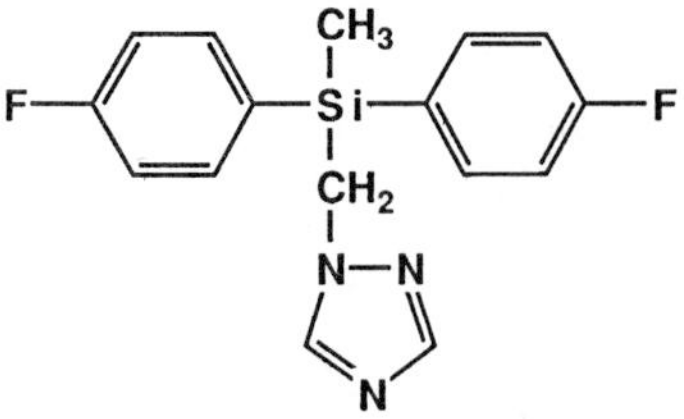

flusilazole
(Du Pont, 1984)

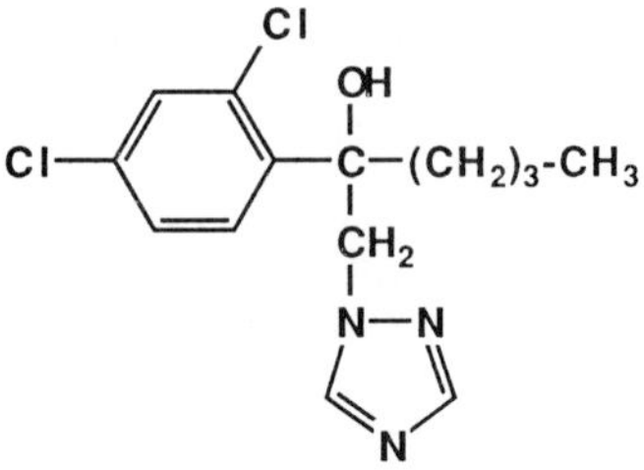

hexaconazole
(ICI, 1986)

Appendix 2:
Specimen product use label

Arrosolo® 3-3E

An Emulsifiable Liquid Herbicide For Use On Rice

An Emulsifiable Liquid Herbicide For Use On Rice

EPA Reg. No. 10182-260

ACTIVE INGREDIENTS:	
S-ethyl hexahydro-1*H*-azepine-1-carbothioate	**33.1%**
N-(3,4-dichlorophenyl) propanamide	**33.1%**
INERT INGREDIENTS:	**33.8%**
	100.0%

Contains 6 Pounds Active Ingredients Per Gallon

Keep Out of Reach of Children

WARNING

Statement of Practical Treatment
Call a Poison Center or Physician Immediately

If a known exposure occurs or is suspected, immediately start the procedures given below and contact a Poison Center, physician or the nearest hospital.

NOTE: Be sure to advise the person contacted that the product is a weak cholinesterase inhibitor in experimental animals. Describe the type and extent of exposure, the victim's symptoms, and follow the advice given.

Note to medical personnel: If cholinesterase inhibition is suspected, atropine by injection is antidotal. It has not been determined if pralidoxime chloride (2-PAM, protopam) is antidotal for the product.

If swallowed: Immediately give several glasses of water but *do not* induce vomiting. If vomiting occurs, give fluids again. Have a physician determine if condition of patient will permit induction of vomiting or evacuation of stomach. Do not give anything by mouth to an unconscious or convulsing person.

If in eyes: Immediately flush eyes with large amounts of water for at least 15 minutes. Hold eyelids apart during flushing to ensure rinsing of the entire surface of the eye and lids with water. Get medical attention immediately.

If on skin: Flush affected areas with plenty of water for several minutes. Remove contaminated clothing and shoes. Get medical attention if irritation occurs.

If inhaled: Remove to fresh air. Seek medical attention if respiratory irritation occurs or breathing becomes difficult.

For 24-hour emergency assistance, call ICI Americas Medical Emergency Information Center 1-800-327-8633. In case of a significant spill, call CHEMTREC 800/424-9300.

FOR OTHER PRODUCT INFORMATION CONTACT YOUR LOCAL ICI AMERICAS INC. DISTRIBUTOR OR SEE PRODUCT SAFETY INFORMATION SHEET.

PRECAUTIONARY STATEMENTS

Hazards To Humans And Domestic Animals

WARNING

Causes substantial but temporary eye irritation. Wear goggles or face shield. Harmful if swallowed. Avoid contact with skin, eyes and clothing. Wash with soap and water after use. Avoid breathing spray mist. Remove contaminated clothing and wash before reuse.

Environmental Hazards: This product is extremely toxic to fish. Do not apply directly to water except as indicated in the directions. Do not contaminate water by cleaning of equipment or disposal of wastes. Do not contaminate water used for domestic purposes.

Physical/Chemical Hazards: Do not use or store near heat or open flame.

General Use Precautions
Read all label directions before using. Do not overdose.

ARROSOLO 3-3E should be used only for recommended purposes and at recommended rates.

ARROSOLO 3-3E may cause crop injury under extreme soil or climatic conditions or if directions are not followed.

Except as specified on this label, do not apply other herbicides, oils, surfactants or liquid fertilizers in conjunction with ARROSOLO 3-3E as crop injury may occur.

Avoid spray drift to all other crops and non-target areas.

Do not apply carbamate or organophosphorus insecticides in combination with ARROSOLO 3-3E.

Do not apply carbamate or organophosphorus insecticides within 7 days before or after ARROSOLO 3-3E application. Severe injury or kill of rice plants may result from tank mix combination or separate sprays of ARROSOLO 3-3E and certain insecticides.

Do not apply ARROSOLO 3-3E when rice is exhibiting stress from very cold temperatures, drought conditions, injury from previously applied pesticides, diseases or insects.

ARROSOLO 3-3E should only be applied to fields which have been drained of flood water to expose susceptible weeds.

To avoid injury do not apply when the air temperature is above 100°F.

Weed control may be impaired if ARROSOLO 3-3E is applied when the air temperature is below 70°F.

For best results apply ARROSOLO 3-3E to actively growing weeds in the 1-2 leaf stage.

Do not apply if rainfall is anticipated within 6 hours after application.

Apply only when wind velocity is low, 0-5 mph.

Do not apply more than 9 pounds actual ORDRAM per acre per season when more than one method of application is used.

General Application Directions
ARROSOLO 3-3E is a selective broad spectrum postemergence rice herbicide for control of annual grasses and broadleaf weeds.

All equipment used in the application of ARROSOLO 3-3E should be carefully calibrated before use and checked frequently during application to be sure the application is uniform. Avoid overlaps that will increase ARROSOLO 3-3E dosage above recommended limits, because crop injury may occur. Apply ARROSOLO 3-3E in 20 to 50 gallons of water per acre by ground equipment and in 10 to 15 gallons of water per acre by airplane.

Shake well before using.

Directions for Use and Recommendations
It is a violation of Federal law to use this product in a manner inconsistent with its labeling.

ICI

Do not apply this product through any type of irrigation system.

POST-EMERGENCE, PRE-FLOOD: For use on dry-seeded and water-seeded rice.

For control of Barnyardgrass, Sprangletop, Dayflower and Broadleaf Signalgrass. Apply ARROSOLO 3-3E at 3 to 4 quarts per acre depending on weed size as shown in the table below. For best results, apply ARROSOLO 3-3E when weeds are in the 1-2 leaf stage. Flood the rice within 5 days after application to prevent weed reinfestation.

Weeds Controlled by ARROSOLO 3-3E and Rate Chart

Weeds		MAXIMUM LEAF STAGE 3 QUARTS	4 QUARTS
Barnyardgrass	*(Echinochloa* spp.)	1-2	3-4
Sprangletop (maximum 1 in. tall)	*(Leptochloa* spp.)	1-2*	1-2
Dayflower, common	*(Commelina communis)*	—	1-2
Signalgrass, broadleaf	*(Brachiaria platyphylla)*	1-2	3-4

*Suppression only in the states of Arkansas, Mississippi, Louisiana and Missouri.

For Bulk and Mini-Bulk Containers

This Label Contains Directions for use on Rice Grown in the Southern Region Only (Arkansas, Louisiana, Mississippi, Missouri, and Texas).

STORAGE AND DISPOSAL

Storage: Do not use or store near heat or open flame. Do not store near seeds and fertilizer. Keep container closed when not in use. Open dumping is prohibited. **Pesticide Disposal:** Wastes resulting from the use of this product may be disposed of on site or at an approved waste disposal facility. **Container Disposal:** Reseal container and offer for reconditioning, or triple rinse (or equivalent) and offer for recycling or reconditioning, or clean in accordance with manufacturer's instructions.

FOR MINI-BULK CONTAINERS

Container Precautions: Before refilling, inspect thoroughly for damage, such as cracks, punctures, bulges, dents, abrasions or worn threads on closure devices. REFILL ONLY WITH ARROSOLO 3-3E. The contents of this container cannot be completely removed by cleaning. Refilling with materials other than ARROSOLO 3-3E will result in contamination and may weaken container. After filling and before transporting, check for leaks. Do not refill or transport damaged or leaking container. CONTAINER IS NOT SAFE FOR FOOD, FEED OR DRINKING WATER!

NOTICE TO BUYER AND USER: Seller warrants that this product conforms to the chemical description on the label and is reasonably fit for the purposes stated on the label when used in accordance with directions under normal conditions of use. This warranty does not extend to the use of this product contrary to label instructions, or under abnormal use conditions, or under conditions not reasonably foreseeable to Seller, and Buyer and User assume the risk of any such use. SELLER DISCLAIMS ALL OTHER WARRANTIES EXPRESSED OR IMPLIED INCLUDING ANY WARRANTY OF FITNESS OR MERCHANTABILITY. SELLER SHALL NOT BE LIABLE FOR CONSEQUENTIAL, SPECIAL OR INDIRECT DAMAGES RESULTING FROM THE USE OR HANDLING OF THIS PRODUCT AND SELLER'S SOLE LIABILITY AND BUYER'S AND USER'S EXCLUSIVE REMEDY SHALL BE LIMITED TO THE REFUND OF THE PURCHASE PRICE.

This product is sold only for uses stated on its label. No express or implied license is granted to use or sell this product under any patent in any country except as specified. Country: United States of America. Patent No.: 3,573,031.

The information and specimen labels in this booklet are correct as of the publication date but are subject to change without notice. For current information, contact ICI Americas Inc.

Bibliography

Alexandratos, N. (1988) *World Agriculture Toward 2000*. FAO/Belhaven Press: London.

Ames, B. N., McCann, J. and Yamasaki, E. (1975) Methods for detecting carcinogens and mutagens with the Salmonella mammalian microsome mutagenicity test. *Mutation Research*, **31**, p. 347.

Blackman, G. E. and Roberts, H. A. (1950) Studies in selective weed control. *Journal of Agricultural Science*, **41**, p. 62.

Bright, J. R. (1968) *Technological Forecasting for Industry and Government*. Prentice-Hall: Englewood Cliffs.

Brown, L. and Eckholm, E. P. (1974) *By Bread Alone*. Praeger: New York.

Carson, R. (1963) *Silent Spring*. Hamish Hamilton: London.

Chou, M., Harmon, D. P. Jnr., Kahn, H. and Wittwer S. H. (1977) *World Food Prospects and Agricultural Potential*. Praeger Publications: New York.

Conner, J. D., Ebner, L. S., O'Connor, C. A., Volz, C. and Weintein, K. (1987) *Pesticide Regulation Handbook*. McKenna, Conner & Cuneo: Executive Enterprises Publications: Washington.

Cook, J. (1977) *Agrochemical Review*, July 1977, **60**, p. 13. Wood MacKenzie: Edinburgh.

Dadd, C. V. (1956) Wild oats; the field problem. *Proceedings of the British Weed Control Conference*, 1956, 1, p. 43.

De Geus, J. G. (1973) *Fertilizer Guide for the Tropics and Sub-tropics*. Centre d'etude de l'Azote: Zurich.

Doane (1965–1985) *Herbicide Market Studies*. Doane Agricultural Services Inc.: St. Louis.

Evans, S. A. (1969) Spraying cereals for control of weeds. *Experimental Husbandry*, **18**, p. 102.

FAO (1978–1988) *Fertilizer Yearbook*. FAO: Rome.

FAO (1978–1988) *Production Yearbook*. FAO: Rome.

Freeman, C. (1974) *The Economics of Industrial Innovation*. Penguin: Harmonsworth.

Fryer, J. and Makepeace, R. (1978) *Weed Control Handbook*. Blackwell: Oxford.

Georghiou, G. P. and Taylor, C. E. (1976) Pesticide resistance as an evolutionary phenomenon. Proceedings of 15th International Congress of Entomology, Washington, p. 759.

Green, M. B. (1976) *Pesticides: Boon or Bane*. Paul Elek: London.

Green, M. B., Hartley, G. S. and West, T. F. (1987) *Chemicals for Crop Improvement and Pest Management*. Pergamon Press: Oxford.

Gunn, D. L. and Stevens, J. G. R. (1976) *Pesticides and Human Welfare*. Oxford University Press: Oxford.

Headley, J. C. (1968) Estimating the productivity of agricultural pesticides. *American Journal of Agricultural Economics,* **50**, p. 13.

IRRI (1974) International Rice Research Institute Annual Report for 1973: Los Banos, Philippines.

Kirby, C. (1980) *The Hormone Weedkillers*. British Crop Protection Council: Croydon.

Large, E. C. (1940) *The Advance of the Fungi*. Jonathan Cape: London.

Leahey, J. P. (1985) *The Pyrethroid Insecticides*. Taylor and Francis: London.

Lever, B. G. (1975) Some economic factors affecting choice of the sterile insect release method as part of a pest control programme: in *Sterility Principle for Insect Control*. International Atomic Energy Agency IAEA — SM, 186/53, p. 135.

Lever, B. G. (1982) The need for plant growth regulators: in *Plant Growth Regulator Potential and Practice*. Ed. Thomas, T. H. British Crop Protection Council: Croydon.

Lever, B. G. (1982) Economics of pest control with emphasis on developing countries: in *Agrochemicals: Fate in Food and the Environment*. International Atomic Energy Agency IAEA — SM, 263/40, p. 41.

Lever, B. G. (1988) The costs of development for a new fungicide: in *Control of Plant Diseases*. Ed. Clifford, B. C. and Lester, E. Blackwell: Oxford.

MAFF (1968) *A Century of Agricultural Statistics*. HMSO: London.

Martin, H. and Woodcock, D. (1983) *The Scientific Principles of Crop Protection*, 7th Edition. Edward Arnold: London.

Matthews, G. A. (1979) *Pesticide Application Methods*. Longman: London.

Peacock, F.C. (1978) *Jealott's Hill: Fifty Years of Agricultural Research*. ICI: Bracknell.

Proctor, J. H. (1974) A review of cotton entomology. *Outlook on Agriculture,* **8**, p. 15.

Reynolds, H. T. (1976) Problems of resistance in pests of field crops. *Proceedings of the 15th International Congress of Entomology*, Washington, p. 794.

Rodemacher, B. (1962) Research and practice of weed control in West Germany. *Proceedings of the British Weed Control Conference*, 1962, **1**, p. 1.

Russell, E. J. (1939) Fertilizers in Modern Agriculture. *MAFF Bulletin,* **28**. HMSO: London.

Russel, E. J. (1956) Weed control: a record of striking progress. *Proceedings of the British Weed Control Conference,* 1956, **1,** p. 3.

Shaw, W. C. (1971) How agricultural chemicals contribute to our current food supplies. *Symposium on Agricultural Chemicals — Harmony or discord for food, people and the environment*. February 1971, Berkeley University, California.

USDA (1978–1988) *Agricultural Statistics*. US Government Printing Office: Washington.

Ware, G. W. (1978) *Pesticides: Theory and Application*. W. H. Freeman: San Francisco.

Watson, J. D., Tooze, J. and Kurty, D. T. (1983) *Recombinant DNA: A Short Course*. W. H. Freeman: New York.

Woodburn, A. and McDougall, J. (1988) *Agrochemical Service*, August 1988. County Nat West Wood Mac: Edinburgh.

Worthing, C. R. and Walker, S. B. (1987) *The Pest Control Manual*. British Crop Protection Council: Thornton Heath.

Index